# Use of Simulation for Training in the U.S. Navy Surface Force

Roland J. Yardley • Harry J. Thie • John F. Schank • Jolene Galegher • Jessie L. Riposo

Prepared for the
**United States Navy**

Approved for public release;
distribution unlimited

# RAND

**NATIONAL DEFENSE RESEARCH INSTITUTE**

The research described in this report was sponsored by the United States Navy. The research was conducted in RAND's National Defense Research Institute, a federally funded research and development center supported by the Office of the Secretary of Defense, the Joint Staff, the unified commands, and the defense agencies under Contract DASW01-01-C-0004.

**Library of Congress Cataloging-in-Publication Data**

Use of simulation for training in the U.S. Navy surface force / Roland J. Yardley ... [et al.].
     p. cm.
     "MR-1770."
     Includes bibliographical references.
     ISBN 0-8330-3481-2 (pbk. : alk paper)
     1. Naval education—United States—Simulation methods. 2. Sailors—Training of—United States.  I. Yardley, Roland J.

VA11.U84 2003
359.5'078—dc22

                                                        2003022269

*Cover photograph: United States Navy photo by Photographer's Mate 1st Class Michael W. Pendergrass (www.news.navy.mil)*

RAND is a nonprofit institution that helps improve policy and decisionmaking through research and analysis. RAND® is a registered trademark. RAND's publications do not necessarily reflect the opinions or policies of its research sponsors.

*Cover design by Stephen Bloodsworth*

Published 2003 by RAND
1700 Main Street, P.O. Box 2138, Santa Monica, CA 90407-2138
1200 South Hayes Street, Arlington, VA 22202-5050
201 North Craig Street, Suite 202, Pittsburgh, PA 15213-1516
RAND URL: http://www.rand.org/
To order RAND documents or to obtain additional information,
contact Distribution Services: Telephone: (310) 451-7002;
Fax: (310) 451-6915; Email: order@rand.org

Navy surface force training has traditionally involved a combination of shore-based and underway training. Recently, however, a combination of economic, operational, and political changes has prompted the Navy to consider shifting the balance toward more shore-based training. The high cost of underway training, the increased operational tempo, reduced access to training ranges, and other factors have decreased the attractiveness of underway training. At the same time, technological advances in simulator technology have made it possible to provide high-quality, shore-based training in many mission areas.

These circumstances prompted the Manpower, Personnel, and Training section of the Assessments Division (N81) of the Deputy Chief of Naval Operations for Resources, Warfare Requirements, and Assessments to ask RAND to examine the potential of using shore-based simulated training to augment live training for the Navy surface force. To address this issue, we conducted a detailed analysis of the training requirements and practices of DDG-51 class ships. In this report, we describe the training architecture for these vessels; specify current requirements, capabilities, and uses of simulation; and analyze the location (under way or in port) of training exercises in relation to training requirements. We also examine the use of simulation for training in other military organizations and the commercial shipping industry. Finally, we present recommendations, based on these analyses, about how simulation and more in-port training might be used to ensure a high level of proficiency in diverse mission areas.

This research was conducted for the U.S. Navy within the Forces and Resources Policy Center of RAND's National Defense Research Institute, a federally funded research and development center sponsored by the Office of the Secretary of Defense, the Joint Staff, the unified commands, and the defense agencies. Comments are welcome and may be addressed to John Schank at John_Schank@rand.org or Harry Thie at Harry_Thie@rand.org. For more information on RAND's Forces and Resources Policy Center, contact the Director, Susan_Everingham@rand.org, 310-393-0411, Extension 7654.

# CONTENTS

# FIGURES

# TABLES

## BACKGROUND

Navy surface force training has traditionally involved a combination of shore-based and underway training. Recently, however, a number of factors—budgetary, political, and environmental concerns, as well as concerns about quality of life for naval personnel—have prompted Navy training officials to consider reducing underway training time and increasing reliance on shore-based simulators. Current personnel practices, such as rotating crews rather than ships to forward-deployed locations, also suggest that requiring crews to complete their training on the ships on which they will be deployed may be impractical. Finally, technological advances have improved productivity and realism in modeling, simulation, and distributed learning. These considerations have given rise to questions about whether relying more heavily on diverse kinds of simulation can decrease underway training and thereby reduce costs but still maintain or improve proficiency and readiness.

## RESEARCH OBJECTIVE AND APPROACH

This research was undertaken to assess the potential of reducing or augmenting underway training by completing more training exercises through simulation. We focused, in particular, on live simulation and virtual, shore-based simulation.

In *live simulation*, a real person operates real equipment, but some aspect of the environment is simulated. Such exercises are carried out on the ship's own equipment—either in port or under way—and are credited toward its readiness rating as specified in the *Surface*

*Force Training Manual* (SURFTRAMAN). Reducing underway time through live simulation would simply involve increasing the proportion of such exercises conducted in port.

In *virtual simulation*, real people operate simulated systems. Reducing underway time through virtual simulation would involve completing more training in shore-based simulators designed specifically for this purpose. Currently, exercises completed through virtual simulation do not count toward a ship's readiness rating, indicating that using simulation to reduce underway time would, ultimately, require changes not only in training procedures but also in training policy.

In this report, we describe

- how simulation and simulators are used for U.S. Navy training for the DDG-51 class ship

- the use of simulation in other military organizations, in civilian aviation, and in commercial shipping

- the relationship between training exercises and the location in which they are conducted

- strategies for increasing the proportion of training exercises completed in port.

## CURRENT SIMULATION POLICIES AND PRACTICES

Currently, the development, governance, financing, and use of simulation is a complex web, with multiple agencies responsible for defining and implementing modeling and simulation (M&S) policy. Furthermore, beyond the basic phase of training, training requirements for ships are only minimally articulated. The vagueness and inconsistency of training requirements and standards for assessing readiness further complicate the problem of determining how simulation might best be used. It would appear that successful use of simulation requires, at a minimum, integrated policy, planning, oversight, management, and adequate resources, as well as a good understanding of benefits and costs. Our review indicates that all these elements are missing, and the subsequent portions of our analysis attempt to provide a part of the empirical foundation needed for more rational policymaking in this domain.

## DDG-51 CLASS TRAINING

The DDG-51 class guided missile destroyer operates offensively in a high density, multithreat environment as a member of a battle group, surface action group, amphibious task force, or underway replenishment group. For this class of ship, the SURFTRAMAN specifies 271 training exercises across 15 mission areas, about 85 percent of which must be repeated at least once within an annual cycle to maintain M-1 training readiness.

By the end of its basic training phase, a ship must have completed 80 percent of these exercises, be proficient (M-2) in all mission areas, have demonstrated the ability to sustain that readiness through its training team organization, and have successfully completed the Final Evaluation Period (FEP) readiness assessment. Subsequently, the ship completes intermediate and advanced training exercises as it prepares to deploy.

Only 58 (21 percent) of the 271 required exercises have approved equivalencies, meaning that credit toward readiness ratings can be earned by completing these exercises either under way or in port. Equivalencies have been approved for exercises in only six mission areas, all of which are tactical. There are no approved equivalencies for exercises in nontactical mission areas.

About one-third of the exercises with the longest time between repetitions (24 months) can currently be simulated, but only about one-tenth of the shortest exercises (three months or less) can be. These findings suggest that simulation has not been applied in the areas that could yield the greatest cost reductions. Because one of the benefits of simulation is being able to repeat training at low cost, it appears that simulation could be most efficiently applied to the exercises that are repeated most often.

In addition, a review of the simulators that can be used for exercises with approved equivalencies for the DDG-51 class, as well as those that are used for all other ships, indicated that a number of simulators support the same exercise. For about 85 percent of exercises that can be simulated, five or more simulators are available for each exercise. This observation suggests that the Navy training community has chosen to simulate better what can already be simulated rather than to explore the application of simulation to different exercises.

## USE OF SIMULATORS IN OTHER ORGANIZATIONS

To determine what issues might arise in efforts to increase the use of simulation for training in the U.S. surface force, we examined how simulation is used and managed in other military and civilian organizations.

In the aviation community, which uses simulators extensively for training, experience indicates that their value depends, among other things, on task characteristics. For instance, a relatively small proportion of training exercises for fighter strike missions is conducted in simulators. For maritime patrol aircraft (MPA), however, about 50 percent of both basic flight and mission training exercises are completed in simulators. In commercial aviation, nearly all training is completed in simulators.

The Canadian Navy requires underway training for only the most advanced exercises—those that involve multiple ships. For all other exercises, the use of simulation is both permitted and encouraged. Among the uses of simulators in the Canadian Navy are familiarizing engineering personnel with machinery control systems, teaching maintenance procedures and ship-handling skills, and training radar navigation teams to use radar displays without the aid of visual references.

In some cases, the simulators the Canadian Navy uses were built as part of the procurement contract for the ships, which permits further cost reductions by integrating simulation in the design of the ship and in plans for training. Some simulators are aboard ships, but others are located at fixed sites, with personnel traveling to them for training. That training can be completed satisfactorily using shore-based simulators at fixed sites indicates that training on a ship's own equipment may be unnecessary for many exercises.

Like the Canadian Navy, the British Royal Navy uses simulation to reduce the costs and risks of training. Underway time is reserved for training in primary warfare mission areas. The Royal Navy has a navigation and ship-handling simulator, as well as a Combat Information Center (CIC) mockup for every class of ship. For Type 22 and 23 ships, the Royal Navy uses these trainers for new officers and also brings CIC teams from existing ships to HMS *Dryad*, one of the Royal Navy's largest training facilities, to run tactical exercises as a prerequisite for operational sea training at Devonport. Operational sea training is used as a postrepair period and providing the Royal Navy's readiness certification process. The combat system suites on the Type 22, 23, and 45 ships all have a simulation or training mode.

In commercial shipping, the International Convention on Standards of Training, Certification and Watchkeeping (STCW) for Seafarers establishes minimum requirements for personnel in certain areas of responsibility, as well as more-general requirements for all crew members, and permits these qualifying tasks to be performed on appropriate simulators. Training officials at the Maritime Institute of Technology and Graduate Studies (MITAGS) indicated that the simulators allow everyone to be trained in a timely manner through a core set of drills. In addition, simulators allow personnel to experience the casualty control exercises and procedures not normally done at sea, thereby preventing equipment damage and personnel injury.

Taken together, these observations indicate that simulation can be and is used effectively to provide many different kinds of training in military and civilian organizations whose needs are comparable to those of the U.S. surface force.

## U.S. NAVY PERSPECTIVES ON SIMULATION

The Navy training representatives we interviewed acknowledged the potential value of using simulation to reduce underway time in surface force training, but they disagreed on how best to use it. Commander, Naval Surface Force, Atlantic (COMNAVSURFLANT) representatives indicated that underway training time could be reduced by completing more intermediate and advanced training in port. Commander Second Fleet (COMSECONDFLT) representatives, on the other hand, suggested that efforts to reduce underway time should focus on basic training. These representatives indicated that, for advanced and intermediate training, ships need to get under way. Although recognizing its potential value, these training officials were concerned that increasing the use of simulation might reduce professional competence, but they did not specify metrics for assessing professional competence, regardless of training method.

In addition to uncertainty about where simulation might be used most fruitfully and general skepticism about assessing its value, there are a number of more-specific factors that might interfere with improved or increased use of simulation. In terms of naval policy, the most important of these factors may be the requirement that any exercise for which completion credit is claimed must be performed on a ship's own systems, i.e., completed through live simulation. This requirement precludes earning "readiness and training" credit for exercises completed in shore-based simulators. To take full advantage of the capabilities of current simulation technology, this requirement would need to be modified.

Another limiting factor that naval officers acknowledged was bias toward completing training exercises under way. A change in Navy culture may be required to achieve a higher level of in-port training involving simulated exercises. Other constraints include the cost of simulators, competing demands on in-port time, and the need to be under way to support other fleet training requirements.

Despite these obstacles, substantial efforts to increase the use of simulation in surface force training have been launched. For instance, COMSECONDFLT was tasked with making in-port integrated warfare training a reality and a standard requirement for battle group interdeployment training. The Battle Group In-Port Exercises (BGIE), with a trial phase conducted from January 2002 through spring 2003, will be conducted as part of BG Interdeployment Training Cycle training. As of this writing, it had not yet been determined whether these exercises will be added to underway exercises or whether they will replace underway training requirements. However, it appears that the Navy has increased its emphasis on in-port training. Further progress in this direction will require not only addressing the concerns noted above but also a detailed understanding of which exercises are now carried out in port and which under way as a basis for identifying exercises for which simulation would be most practical and beneficial.

## WHERE DDG-51 TRAINING EXERCISES ARE COMPLETED NOW

In terms of possible completion sites, there are three classes of DDG-51 training exercises:

- exercises with equivalencies
- exercises with no equivalencies that can be completed under way or in port

- exercises with no equivalencies that can only be completed under way.

To assess the relationship between where exercises *could* be completed and where they *were* completed, we conducted a detailed quantitative analysis relying on four kinds of information: the locations of ships on specific dates during the period under study; the exercises that were completed and the dates on which they were completed; whether or not each of the completed exercises had an approved equivalency; and whether, in the views of Naval Surface Force, Atlantic (SURFLANT) training officials, exercises with no equivalencies could be completed under way or in port or could only be completed under way.

Table S.1 summarizes this analysis. Of the 8,250 exercises completed by nondeployed ships, 6,356 exercises (77 percent) were completed under way, and 1,894 (23 percent) were completed in port. Of exercises with equivalencies, all of which could be completed in port, 80 percent were completed under way. Only 20 percent were reported as having been completed in port. Thus, the existence of an approved equivalency did not affect the likelihood that an exercise would be completed under way. Neither did mission area appear to determine whether these exercises are completed in port or under way. Most exercises with equivalencies were completed under way, regardless of mission area. Further, a large majority (71 percent) of

**Table S.1**

**All Exercises Were Completed**

|  | Exercised Completed | |
| --- | --- | --- |
| Exercise Type | Under Way (%) | In Port (%) |
| All exercises[a] | 77 | 23 |
| Exercises with equivalencies | 80 | 20 |
| Exercises that could be completed under way or in port | 71 | 29 |
| Exercises that could only be completed under way | 92 | 8 |

[a]All figures for nondeployed ships only. For deployed ships, 99 percent of all exercises were completed under way.

the exercises that could, according to SURFLANT training officials, be completed under way or in port were also completed under way. Only 29 percent were completed in port.

Table S.1 also indicates that 8 percent of "underway only" exercises were completed in port. This finding may be a result of reporting error, or it may suggest that exercises that were categorized as "must be completed under way" are, in fact, sometimes completed in port.

Table S.2 summarizes the results of our analyses for exercises that could be completed in port that actually were completed in port. The data in the first column reflect 6,756 potential in-port exercises completed during the period we examined. Of these, 21 percent had equivalencies, and 79 percent were exercises that did not have equivalencies but that could have been completed either under way or in port. The second column shows the percentage of each type of potential in-port exercise that actually was completed in port. Exercises with equivalencies actually completed in port constitute 4 percent of exercises that could have been completed in port; exercises with no equivalencies that could be completed either in port or under way and were actually completed in port constitute 22 percent of potential in-port exercises. In sum, only 26 percent of all potential in-port exercises were actually completed in port—about one-fourth of the exercises that could have been completed in port. These data clearly indicate that there is substantial opportunity for increasing the proportion of training exercises conducted in port and, by extension, for increasing the use of simulation in training.

### Table S.2

### Proportion of All Exercises That Could Be Completed in Port Actually Completed in Port

| Exercise Type | Exercises That Could Be Completed in Port | |
|---|---|---|
| | Possible (%) | Actual (%) |
| Exercises with equivalencies | 21 | 4 |
| Exercises with no equivalencies that could be completed in port or under way | 79 | 22 |
| Total | 100 | 26 |

## RECOMMENDATIONS

Based on the analyses described above, we developed several recommendations regarding the use of simulation for training in the surface force. In presenting these recommendations, our aim is to provide information that can help to

- make optimal use of underway training time
- guide decisions about whether to increase the use of virtual and live simulation in surface force training
- identify mission areas and kinds of exercises that would be appropriate targets for increasing the use of simulation, especially virtual simulation
- develop strategies for purchasing and implementing simulators.

### Define the Goals of the Training

To determine the roles of live and simulated training in relation to performance goals, it is essential to establish the goal of training. If the goal is to achieve the greatest proficiency, more resources have to be expended or significant process changes need to be made. If the goal is to reduce cost while maintaining the same proficiency, the trade-offs may be different. Defining clear training goals could help to increase openness to the use of simulation for training.

### Specify Measures of Effectiveness for Training

The Navy needs clear proficiency and readiness standards covering all phases of training across mission areas to assess not only the general effectiveness of training but also the efficacy of training through simulation. Given the experience of other organizations, it seems likely that live in-port or virtual simulation would be at least as effective as underway training for many skills. Evidence derived from well-defined measures substantiating such an outcome could be instrumental in producing change.

### Increase the Efficiency of Underway Training

Underway training time should be reserved for exercises that can only be completed under way. The number of exercises with approved equivalencies should be expanded, and a much-higher proportion of exercises with equivalencies should be completed in port. In addition, "underway only" exercises should be prioritized, with high-priority exercises being completed first. Finally, exercises should be sorted into groups that can be completed simultaneously, so as to maximize the training benefit from time at sea.

## Develop a Simulation Strategy

To expand the use of simulation in a way that optimizes both proficiency and the use of training resources, we propose a three-pronged strategy. First, the Navy needs to clarify responsibility and authority for decisionmaking with regard to the use of simulation in training. Second, earning credit through training on shore-based simulators must be made permissible under the SURFTRAMAN. Third, the Navy should select areas for simulation in which simulation will provide the greatest benefit. In the following paragraphs, we identify several categories of such exercises.

**Exercises for Which the Actions Taken Do Not Depend on the Location of the Ship.** Exercises for which it is known that the actions taken or reactions to stimuli do not depend on the location of the ship should be regarded as candidates for simulation. For example, an antisubmarine warfare team can accomplish approximately the same level of training when the ship is tied to the pier, with external signals stimulating equipment, as it can under way.

**High-Frequency Exercises.** Logically, it seems that exercises that require repetition would be good candidates for simulation because the consistency of the training environment would allow users to develop and refine their skills without the intrusion of irrelevant factors that may undermine performance. It also seems likely that simulating high-frequency exercises would be more cost-effective than repeating exercises while under way, and the financial benefits of simulating high-frequency exercises could be maximized by focusing development efforts on the most-costly high-frequency exercises.

**Exercises in Nontactical Mission Areas.** Currently, all approved equivalencies are in tactical mission areas, but training for many nontactical missions could take place in port. The use of simulation by other militaries and the private sector indicates that engineering, ship-handling functions, and maintenance exercises are all likely candidates for simulation.

**Exercises Simulated by Other Military Organizations and the Private Sector.** As we have shown, other military organizations, commercial aviation, and commercial shipping use simulation for training to a much greater extent and with good results than does the U.S. Navy. The Navy is currently moving to align its qualification standards with those of the merchant marine. As this process goes forward, it should focus not only on standards but also on methods of meeting those standards. The use of simulation for training should be given a key role in defining these methods.

## Develop an Investment Strategy

The Navy should invest in the simulators that afford the best fidelity and maximize their availability. Heretofore, limited availability of simulators with good fidelity has hindered the expansion of training through simulation.

Because there are multiple simulators for the same exercises, it appears that the direction of simulation has been to improve the fidelity of what has already been simulated. Although fidelity is, of course, important, fidelity may already be satisfactory in some areas, making it desirable to give greater attention to developing simulators for new mission areas.

# ACKNOWLEDGMENTS

The project team would like to thank the staff of the Navy N81 office for their support, especially Richard Robbins, CDR Robert Henderson, CDR Daniel Boyles, and Douglas Corum. We also thank the individuals who assisted us in data collection, as well as those who shared their insights and expertise regarding naval training. In particular, CAPT William Valentine, COMNAVSURFLANT, N7, and CDR Darrel Morben, COMSECONDFLT, J7B, met with us to explain the Navy training process; the information they provided greatly enhanced our understanding of fleet training and simulation. In addition, many officers in the Navy training community helped us understand fleet training requirements and how they are achieved. The staff of the Maritime Institute of Technology and Graduate Studies provided a detailed description of the use of simulation in support of training requirements for merchant mariners. CDR Laurence Smallman and CDR Paul Casson of the United Kingdom embassy and CAPT Peter Hoes and LCDR Robert Craig of the Canadian embassy provided details into the role simulation plays within their respective navies. We are also grateful to our RAND colleagues Kevin Brancato, who supported us in merging and evaluating the database; John Bondanella and Peter Schirmer, who provided many valuable suggestions on an earlier version of this report; and Phyllis Gilmore, who edited the result.

The views expressed herein are our own and do not represent the policy of the Department of the Navy.

| | |
|---|---|
| AAW | antiair warfare |
| ACTS | Aegis Combat Trainer System |
| AMW | amphibious warfare |
| ARG | Amphibious Ready Group |
| ASU | Administrative Support Unit |
| ASW | antisubmarine warfare |
| ATG | Afloat Training Group |
| AWW | above water warfare (Canada) |
| BEWT | Battle Force Electronic Warfare Trainer |
| BFTT | Battle Force Tactical Training |
| BGIE | Battle Group In-Port Exercises |
| C2W | Command and control warfare |
| CART | Command Assessment of Readiness and Training |
| CCC | command, communication, and control |
| CFFC | Commander, Fleet Forces Command |
| CIC | Combat Information Center |
| CIO | Chief Information Officer |
| CMTpc | Cruise Missile Trainer Personal Computer |
| CNET | Chief of Naval Education and Training |
| COMFLTFORCOM | Commander, Fleet Forces Command |

| | |
|---|---|
| COMLANTFLT | Commander, U.S. Atlantic Fleet |
| COMNAVSURFLANT | Commander, Naval Surface Force, Atlantic |
| COMOPTEVFOR | Command Operational Test and Evaluation Force |
| COMSECONDFLT | Commander Second Fleet |
| COVE | Conning Officer Virtual Environment |
| CRR (Canada) | combat readiness requirements |
| CVBG | carrier battle group |
| CWC | Composite Warfare Commander |
| CY | calendar year |
| DoD | Department of Defense |
| DoDD | Department of Defense Directive |
| DON | Department of the Navy |
| EGCS | Environmental Generation & Control System |
| ENWGS | Enhanced Naval Wargaming System |
| EW | electronic warfare |
| FAM | Functional Area Manager |
| FCA | Fleet concentration area |
| FEP | Final Evaluation Period |
| FXP | Fleet Exercise Publication |
| HMS | Her Majesty's Ship |
| HQ | headquarters |
| IDTC | Interdeployment Training Cycle |
| IMPASS | Integrated Maritime Portable Acoustic Scoring and Simulator |
| INT | intelligence |
| ISIC | Immediate Superior in Command |
| ITS/TCD | Interim Trainer Support/Trainer Control Device |
| LAMPS | Light Airborne Multipurpose System |

| | |
|---|---|
| LANTFLT | Atlantic Fleet |
| M&S | modeling and simulation |
| MITAGS | Maritime Institute of Technology and Graduate Studies |
| MIW | mine warfare |
| MOB-D | Mobility–Damage Control |
| MOB-E | Mobility–Engineering |
| MOB-N | Mobility–Navigation |
| MOB-S | Mobility–Seamanship |
| MPA | maritime patrol aircraft |
| MUTTS | Multiunit tactical training system |
| NATO | North Atlantic Treaty Organization |
| NAVAIR | Naval Air Systems Command |
| NAVMSMO | Navy Modeling and Simulation Management Office |
| NCO | noncombat operations |
| NSFS | Naval Support Fire Support |
| OPNAV | Chief of Naval Operations |
| POE | Projected operational environment |
| ROC | Required operational capability |
| S&T | science and technology |
| SECNAVINST | Secretary of the Navy Instructions |
| SG&R | scenario generation and reconstruction |
| SH-60B | Seahawk Helicopter |
| SORTS | Status of Resources and Training System |
| SPAWAR | Space and Naval Warfare Systems Command |
| SSAAC | Surface Ship Acoustic Analysis Center |
| STCW | Standards of Training, Certification, and Watchkeeping for Seafarers |
| STW | strike warfare |

| | |
|---|---|
| SURFTRAMAN | Surface Force Training Manual |
| SUW | surface warfare |
| SWO | Staff Warfare Officer |
| SWOS | Surface Warfare Officer School |
| TACDEW | Tactical Advanced Combat Direction & Electronic Warfare |
| TCD | Trainer Control Device |
| TRMS | Training Readiness Management System |
| TRNGREP | training report |
| TSTA | Tailored Ship Training Availability |
| USAF | United States Air Force |
| USMC | United States Marine Corps |
| USN | United States Navy |
| USW | Undersea warfare |
| VAST | Virtual At-Sea Training |

# Chapter One

# INTRODUCTION

## BACKGROUND

The Navy training continuum consists of individual, team, unit, multiunit, and battle group training. Navy surface force training, particularly at the team and unit levels, has traditionally focused on a combination of shore-based and underway (at-sea) training. Recently, however, increasing budget and political pressures have reduced access to training ranges, thereby constraining underway training. Also, increased operational commitments have reduced the time that ships spend in port, adversely affecting the time available for sailors to spend with their families and, therefore, the quality of life of fleet personnel. Finally, environmental concerns constrain the locations in which underway exercises can be completed, and the geographic dispersion of ships may mean that some ships will have to sail great distances to complete group training. Taken together, these factors suggest that it may be desirable to reduce underway training requirements and increase reliance on shore-based simulators.

In addition to these incentives to reduce underway days, current personnel practices also suggest that increasing the use of simulation in training may be useful. For example, the Navy has been experimenting with rotating crews, rather than ships, to forward-deployed locations. This new operating procedure saves fuel and transit time and increases time on station for the unit that is forward deployed. It also requires naval personnel to be resourceful and adaptable enough to learn about and operate equipment that is configured differently on the ships that they deploy to from the way it is on the ships on which they trained. Simulation could be used to help achieve this flexibility.

As these changes have occurred, technological advances have improved productivity and realism in modeling, simulation, and distributed learning, but these improved technologies have not influenced the Navy's training program. The relationships between live, underway training, and in-port training with simulators have not changed significantly over the last few decades. Given these advances in simulation capabilities, questions arise about the extent to which the use of simulators could reduce the need for underway training, thereby reducing costs, while maintaining or increasing proficiency and readiness.

Because of the complex training requirements and operating procedures for surface ships, there is no straightforward answer to such questions. The crews on naval vessels range from fewer than 100 to several thousand individuals, who have dozens of ratings and responsibilities. These sailors must work together to control and protect the ship while conducting a wide range of operational missions. The training requirements for their diverse tasks vary widely.

Choosing the best approach to training for a particular task requires not only understanding the relationship between task requirements and the cost and effectiveness of particular training methods but also understanding current and potential links between training exercises. When a ship leaves port for an at-sea training evolution, it conducts as many training events as possible over various mission areas. Some of these events are at the individual and unit levels, while others are designed to improve coordination and integration of all the individuals and teams on the ship. The variation in kinds and levels of training exercises highlights the possibility of linking training for diverse tasks sequentially or hierarchically, with individual training embedded in unit or battle group training or training for individual tasks embedded in exercises designed to meet broader mission requirements.

How such linkages might affect requirements for underway training, however, is unclear. Substituting shore-based simulation for underway training may be effective for a subset of individuals or operational missions, but, to the extent that underway training is needed for other tasks, increasing the use of shore-based simulation could increase costs without reducing the necessity for underway training. Thus, an analysis of where and how simulations of various kinds are or might be used is clearly warranted.

## RESEARCH OBJECTIVE AND APPROACH

This research was undertaken to assess the potential of using simulated, in-port training to replace or augment underway training for the U.S. Navy surface force. The Manpower, Personnel, and Training section of the Assessments Division (N81) of the Deputy Chief of Naval Operations for Resources, Warfare Requirements, and Assessments sponsored the project. Our goals were to describe

- how simulation and simulators are used for U.S. Navy training for the DDG-51 class ship
- the use of simulation in other military organizations, in civilian aviation, and in commercial shipping
- the relationship between training exercises and the location in which they are conducted
- strategies for increasing the proportion of training exercises completed in port.

To achieve these goals, we carried out a multifaceted investigation, involving both qualitative and quantitative methods, following the steps described below.

### Step 1. Describe the Training Environment and the Use of Simulation

We first developed a detailed description of the current organizations, processes, and policies for both live and simulated training, as well as plans for the future use of simulation technologies. This description constitutes an overview of surface force training with regard to simulation.

### Step 2. Describe Training Requirements for DDG-51

To link this overview to a specific context, we extracted a detailed description of training requirements for the DDG-51 class ship from the *Surface Force Training Manual* (SURFTRAMAN).[1] This descrip-

---

[1] COMNAVSURFLANT, 1999, p. 1-1-1. However, unless otherwise indicated, as here, references to the SURFTRAMAN in this report are to the 2002 edition.

tion includes the set of mission areas for the DDG-51, the training courses that must be completed, and the exercises associated with each course. In addition, we identified exercises for which there are approved equivalencies and exercises that require repetition. We also described the simulators that are currently used for training on this class of ship and the procedures that are used for assessing readiness.

## Step 3. Describe Simulator Use in Other Military Organizations, Civilian Aviation, and Commercial Shipping

To complete this step, we interviewed naval personnel from the United Kingdom and Canada, as well as staff from the Maritime Institute of Technology. We also drew on our previous research on the use of simulation in the aviation community (both civilian and military) to determine how the issues encountered in the use of simulation in those contexts might inform the development and implementation of simulators for training in the Navy's surface force (Schank et al., 2002).

## Step 4. Link Location of Ship to Exercise Completion

Using material we obtained from training reports and ship employment schedules, we specified which exercises had been completed by DDG-51 class ships and the locations of all ships in the class for each date during the period under study. We then interviewed training personnel from the type commanders and numbered fleet commanders who are responsible for defining training requirements and for conducting the required training. Relying on the results of our interviews, we classified exercises that did not have equivalencies as "exercises that could only be completed under way" or "exercises that could be completed under way or in port." These observations were then combined in a database that allowed us to link the location of ships to exercise completion for each type of exercise.

## Step 5. Specify Type and Proportion of Exercises Completed Under Way Rather Than in Port

Using the database described above, we examined the relationship between where exercises could be completed—based on the requirements specified in the SURFTRAMAN and the views of the training officials we interviewed—and where they were completed. This analysis enabled us to specify the type and proportion of exercises completed under way and in port.

## Step 6. Analyze Change Possibilities and Develop Recommendations

Based on the results of our quantitative and qualitative analyses, we identified several possibilities for changing training policies and practices. These possibilities take into account practical considerations, such as the competing demands on the availability of personnel to participate in in-port training as opposed to other in-port requirements and training goals. We concluded our analysis by developing a set of recommendations for changes in training policies and practices, including areas in which the use of in-port training could be increased.

## DEFINING THE RESEARCH ENVIRONMENT

In reality, all training involves some type of simulation. Live training typically signifies the actual operations of the equipment, either an aircraft or a ship, during training events. Even when training is live, other aspects of the event—such as the environment, weapons, targets, or threats—are typically simulated. Throughout this report, we use the terms *live, under way,* and *at sea* to refer to ships operating away from their home ports. Such ships could be deployed to an operational area or training for future deployments.

*Simulation* is defined as a technique for testing, analysis, or training in which real-world systems are used or in which a model reproduces real-world and conceptual systems (DoD, 1994). One widely used taxonomy defines three types of simulation: live, virtual, and constructive.

## Live Simulation

In live simulation, a real person is operating real equipment, but some aspect of the environment or another parameter is simulated. An example of live simulation is the Aegis Combat Trainer System (ACTS), in which Combat Information Center (CIC) personnel operate their own equipment onboard a DDG-51. In this situation, naval personnel use real equipment in a simulated operational environment.

As we will discuss, live simulation is the only type of simulation that can be used for completing some training events. As our findings will indicate, however, almost all training events for the surface force are accomplished through live simulation even when live simulation is not clearly required. Thus, one aspect of our analysis focuses on reducing underway days by increasing the use of live simulation while a ship is in port.

## Virtual Simulation

In virtual simulation, real people operate simulated systems. For example, virtual simulations allow naval personnel to exercise motor-control skills (e.g., flying an airplane), decision skills (e.g., committing fire-control resources to action), and communication skills (e.g., exchanging information among members of a command, control, communications, computers, and intelligence team) (DoD, 1995). The term *simulators* typically designates these simulated systems or equipment.

By its very nature, virtual simulation in surface training is accomplished while the ship is not under way. A second aspect of our research focuses on increasing the use of virtual simulation for surface force training. This approach would require changing policies to grant training credit for training events that are not completed on the ship's equipment (i.e., through live simulation) but instead involve sending individuals or teams to shore-based training simulators.

## Constructive Simulation

In constructive simulation, both the personnel and the system they are operating are simulated. Real people take actions within (i.e., cause a response in) this environment, but the effects of their actions are determined by parameters built into the simulator. That is, the simulator provides feedback to the user about the effects of his or her decisions regarding the actions of simulated people within a simulated environment. The user's decisions are inputs to the environment, and the results of these decisions are determined by the parameters built into the program. Computer-based war-gaming is an example of constructive simulation.

Although useful, this taxonomy is somewhat problematic. First, its categories have fuzzy boundaries; simulators cannot always be described as being clearly of one of these three types. The degree of human participation in the simulation is infinitely variable, as is the degree of operational realism. Second, the taxonomy excludes a fourth category: simulated people operating real equipment, such as smart vehicles (Naval Research Advisory Committee, 1994). Thus, the taxonomy should not be seen as a precise description of the diverse simulators that exist now or might be built. Nonetheless, it has heuristic value in that it directs attention to the diverse ways in which aspects of a training exercise might be simulated.

The main question we address is not whether a higher degree of simulation or more simulators can be used in surface force training but rather whether more training can be accomplished in port than under way. Achieving that goal, however, may require more simulations and simulators.

## ORGANIZATION OF THIS DOCUMENT

Chapter Two describes the training cycle for Navy ships and the organizations, policies, and procedures that govern the use of simulation in training. It then describes training for the DDG-51 class of ships and how simulators can be used to accomplish training events. Chapter Three discusses how other organizations—including the Canadian and British navies, the aviation community, and commercial shipping—use simulators for training. Chapter Four summarizes the results of a literature review and discussions with various Navy organizations regarding the current use of and future directions for simulation in surface force training. In Chapter Five, we present our analysis of where training events are completed, either under way or in port, and identify possibilities for performing more training in port. Chapter Six summarizes our findings and provides recommendations.

# Chapter Two

# OVERVIEW OF THE USE OF SIMULATION
# IN SURFACE FORCE TRAINING

In this chapter, we describe the Interdeployment Training Cycle (IDTC), during which the training that concerns us takes place. We also describe the SURFTRAMAN as the source of training policy and procedures for the U.S. Navy's surface force. In addition, we specify the components of a training architecture with respect to simulation and illustrate the complexity of policies governing the use of simulation in training. This discussion provides the background for our description of training requirements for the DDG-51 class of ships.

## THE INTERDEPLOYMENT TRAINING CYCLE

The IDTC—the time between major deployments—is usually about 18 months. This cycle involves a number of training events and evaluations that occur in three distinct phases: basic, intermediate, and advanced. Figure 2.1 provides an overview of the IDTC.

## Command Assessment of Readiness and Training

Two important assessments of a ship's readiness take place before IDTC begins: Command Assessment of Readiness and Training (CART) I and II. The results of these assessments are used to specify the training required to ensure mission readiness.

The CART I is conducted toward the end of a ship's deployment; it is a ship's self-assessment of projected personnel turnover, required school graduates, and personnel rotation and succession plans for the next deployment.

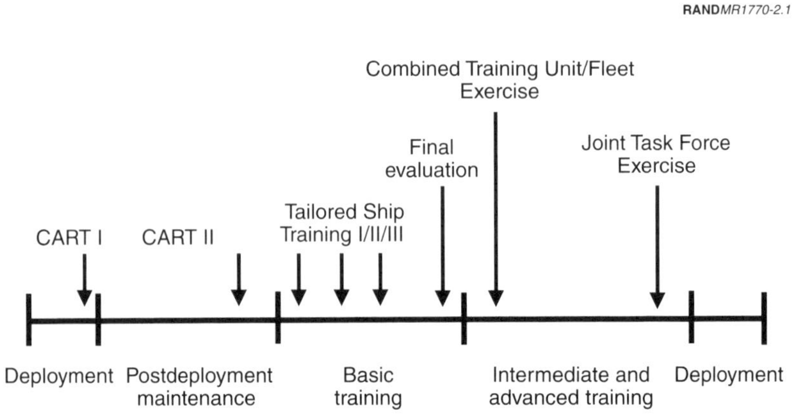

RAND*MR1770-2.1*

Figure 2.1—The Interdeployment Training Cycle

At the end of the deployment, a ship will complete a maintenance period. Near the end of the postdeployment maintenance period, the CART II is conducted. In that assessment, the ship's Immediate Superior in Command (ISIC), assisted by the Afloat Training Group (ATG), evaluates the ship. This evaluation is used to identify training deficiencies and to tailor training plans to meet these deficiencies.

## Training Phases in the IDTC

Once the assessments are complete, training progresses through three phases: basic, intermediate, and advanced. During these phases, naval personnel engage in training exercises of increasing complexity, focusing first on individual and unit-level training and, eventually, shifting their attention to training to serve as one unit within a battle group. When not deployed, ships are allocated 28 steaming days per quarter to accomplish training and to meet other commitments, such as port calls, or to carry out supporting missions, such as drug interdiction.

Basic training focuses on unit-level training emphasizing basic command and control, weapon employment, mobility (navigation, seamanship, damage control, engineering, and flight operations), and warfare specialty. A key objective during this period is satisfactory completion of required certifications. Upon completing the basic phase, a unit is expected to be substantially ready (M-2) in all mission areas.[1]

Throughout basic training, a ship will undergo Tailored Ship's Training Availability (TSTA), during which it completes a combination of in-port and underway exercises assisted by ATG personnel. These exercises are designed to address the specific training needs that were identified during the CART I and II. The end of basic training is marked by a Final Evaluation Period (FEP), during which the ATG assists the ISIC in assessing a ship's readiness to move beyond basic to intermediate training.

Intermediate training focuses on warfare team training in support of the Composite Warfare Commander (CWC) organization. It is conducted ashore under tactical training groups and at sea under the training carrier or amphibious group commander, culminating in a Composite Training Unit Exercise. In intermediate training, ships will work in their primary and secondary warfare areas with one or more units. During this phase, ships begin to develop warfare skills in coordination with other units while continuing to maintain unit proficiency.

Advanced training focuses on coordinated battle group warfare skills. The numbered fleet commanders conduct this phase, which includes shore-based war-gaming. The at-sea phase is devoted to a fleet exercise that evaluates all warfare skills.

As this description suggests, a ship's readiness level is based on successful completion of a large number of training exercises. Some of these exercises must be repeated periodically to maintain readiness, and these are conducted during all three training phases and when a ship is deployed.

## TRAINING POLICY AND GUIDANCE

The SURFTRAMAN is the primary source for type commander training policy and training requirements, specifying the minimum requirements for a program that integrates individual, team, and unit training. It is the primary directive for planning, scheduling, and exe-

---

[1]*M-ratings* are time-phased indications of degradation of proficiency. *M-1* describes the state of readiness immediately after exercise completion; after a specified period, the readiness for the particular exercise would degrade to *M-2*, and so forth.

cuting all training requirements within the naval surface forces (COMNAVSURFLANT, 1999). The manual is issued by the commander of the naval surface forces for use by the Atlantic and Pacific fleets and is updated periodically as needed.

The guidance in the SURFTRAMAN includes

- requirements to maintain basic phase proficiency throughout the IDTC
- skills that must be practiced to maintain proficiency and readiness
- equivalencies, i.e., specification of exercises that can be accomplished through constructive simulation, either in port or under way.

A ship's mission is performed under way; thus, almost all exercises *can* be completed under way. Many exercises, however, do not *require* that a ship be under way to complete them or to take credit for their completion. Exercises that can be and are completed in port are counted toward a ship's mission readiness in the same way as exercises that are completed under way.

The SURFTRAMAN (p. 4-1-1) provides the following guidance for the conduct of in-port training:

> Inport training provides a controlled test of systems and equipment as well as a method to verify personnel assignments before going to sea after an extended inport period. Additionally, it provides an alternative to underway exercise periods for completion of many types of training. Exercises conducted on board can encompass a wide spectrum of operational training including equipment operation, watch standing, watch team procedures, and tactics.

Appendix C of the SURFTRAMAN provides a matrix of exercises that may be completed by scenarios generated from shore-based, mobile, or embedded generators—but run on the ship's own systems—that are approved for readiness reporting. We have listed these exercises in Appendix A of this report.

During our visit to Commander, Fleet Forces Command (CFFC), CFFC training representatives indicated that simulation is valuable as a foundational element of training in that it enables crew members to set up, start, operate, and become familiar with equipment.

In addition to developing technical proficiency, these training exercises are designed to help sailors establish and maintain the internal and external communication systems needed for connectivity and mission readiness. Working through these challenges in a simulated environment improves training and better prepares teams to address similar challenges under way.

Because the ability to simulate an at-sea environment is critical to the success of structured in-port training exercises, these exercises must include consideration of the availability of appropriate training devices and coordinated planning between the officer in charge of the exercise and participating units.

## SIMULATION POLICIES AND PRACTICES

A number of challenges arise in trying to assemble an overview of simulation policies and practices. The biggest challenge stems from the combinatorial size of the problem. There are 38 organizations that manage tactical training and 39 that provide tactical training, 60 schoolhouses, at least 11 resource sponsors, at least ten different publications that specify policy for modeling and simulation (M&S), and an unknown number of simulators. By 2008, there will also be 56 DDG-class ships, the class we discuss here, each of which has different simulation technologies and capabilities.

Beyond the basic phase of training, training requirements for these ships are only minimally articulated. Each ship must complete a set of training exercises that has been tailored to counter the ship's deficiencies. These exercises are selected and planned by a number of different people. In the intermediate and advanced phases of training, there is little consistency in the choice of exercises or the requirements that must be met, and, in all phases of the IDTC, a number of different persons evaluate the success of training. The vagueness and inconsistency of training requirements and standards for assessing readiness further complicate the problem of determining how simulation might best be used.

Several organizations, including the Department of Defense (DoD), the Secretary of the Navy, and the Fleet Forces Command Operational Test and Evaluation Force (COMOPTEVFOR), have promulgated M&S policy. Table 2.1 lists some of the directives that address simulators in training.

**Table 2.1**

**Chronology of Directives Regarding the Use of Simulators**

| | |
|---|---|
| DoD Directive 1430.13 | August 1986 |
| DoD Directive 5000.59 | January 1994 |
| COMOPTEVFORINST 5000.1 | September 1995 |
| DoD Directive 5000.59-P | October 1995 |
| DoD Directive 5000.61 | April 1996 |
| Navy M&S Master Plan | February 1997 |
| SECNAVINST 5200.40 | April 1999 |
| SECNAVINST 5200.38 | February 2002 |
| CFFINST 3502.1 | March 2002 |

Early directives authorized DoD to use simulators to make training more effective and to help maintain military readiness (ASD FM&P, 1986). Training simulators permit

- training that might be impractical or unsafe if done with actual systems or equipment

- concentrated practice in selected normal and emergency actions

- training of operators and maintainers to diagnose and address possible equipment faults

- increases in proficiency despite shortages of equipment, space, ranges, or time

- control of life-cycle training costs

- reduction of systems required in maintenance training.

Similar themes appeared in M&S policy that followed and in the most current guidance (COMFLTFORCOM March 2002a), which established an M&S fleet project team to ensure that synthetic training systems are developed in a coordinated manner.

Whether such coordination can be achieved remains to be seen, given the number of organizations currently involved in some aspect of the development, governance, and application of simulation in training. (See Table 2.2.) These organizations have not updated their policies and guidelines to reflect new processes and technology, and the policies and guidelines are often inconsistent with each other.

## Table 2.2

## Naval Organizations and Their Responsibilities with Regard to Simulation

| | |
|---|---|
| Under Secretary of the Navy | Senior official in DoN for M&S |
| M&S Advisory Council | Oversight for DoN M&S |
| N6 | Navy M&S Executive Agent, co-chair M&S advisory council |
| NAVMSMO | Navy M&S single point of contact; develop policy, strategy, plans, and directives; establish supporting organizations, build consensus for investment strategy |
| M&S Technical Support Group | Provide technical advice, define standards, assess ongoing government and industry efforts, maintain Navy M&S Resource Repository, manage verification, validation, and accreditation process |
| M&S Program Office | Manage development of synthetic battlespace, manage building block programs, integrate components, supported by S&T effort to assess emerging technologies |
| M&S S&T Center | Conduct research underpinning Navy M&S |
| Functional Area Manager (FAM ) | Coordinate functional area, represent Navy on DoD and Joint Staff M&S working and sub–working groups and Task Forces through NAVMSMO |
| Integrated Planning Team | Participate in development of M&S policies, procedures, guidelines, and plans; review, prioritize, and recommend opportunities for joint or collaborative M&S development; review and recommend M&S sponsors and proponents |
| Resource Sponsors | Support M&S strategy, validate requirements, fund M&S through program objectives memorandum process, sponsor M&S applications, quantify impact of M&S |
| M&S Executive Agents | As delineated in DoD 5000.59-P M&S Master Plan |
| M&S Developers & Users | Establish needs and requirements, register models and data, manage configuration of M&S applications, quantify impact, build and maintain M&S in accord with standard, verify and validate components |

The organizational structure and jurisdictions that surround simulation with respect to training are overly complex. A large number of decisionmakers and their various offices have overlapping responsibilities, and each new policy directive and each staff realignment add more layers of review and decisionmaking because the older forms are not eliminated. As a result, it is difficult to determine who is in charge of issuing policies and orders and who is responsible for following them.

Given the complexities of the uses of simulation that we have presented, as well as the complexity of the more-general training environment, we have constrained our analysis in several ways:

- We focus on one class of ship: the DDG-51.

- We are concerned with simulation as it is used to train teams or units, not the use of simulators to train individuals in a school-house.

- We treat the training requirements specified in the SURFTRA-MAN as the basis of our analysis of how simulation is used and how it might be used.

- We treat mission area as the focus of training and the realm in which simulation is described.

## DDG-51 CLASS TRAINING

The mission of the DDG-51 class guided missile destroyer is to operate offensively in a high density, multithreat environment as an integral member of a battle group, Surface Action Group, Amphibious Task Force, or Underway Replenishment Group to include striking targets along hostile shorelines (OPNAV, 1994).

Appendix A of the SURFTRAMAN delineates required training exercises by mission area, as well as in-port training drills and other training events. For the DDG-51 class ship, SURFTRAMAN specifies 271 exercises across 15 mission areas. Table 2.3 shows the number of exercises in each mission area.[2] These are Fleet Exercise Publication

_____

[2]Appendix A provides information on the training events in each mission area including the titles of the events, their periodicities, whether or not they have equivalencies,

(FXP) exercises that directly determine the ship's Status of Resources and Training System (SORTS) mission area readiness (OPNAV, 1997b). They are designed to support the training of units in each of their warfare mission areas, ensuring operational capabilities in each projected operating environment.

A ship must progress through completion of these exercises so that, by the end of the basic phase, it has completed 80 percent of the FXP exercises, is proficient (M-2) in all mission areas, has demonstrated the ability to sustain that readiness through its training team organization, and has successfully completed the FEP readiness assessment (COMNAVSURFLANT, 1999, p. 1-1-1). Subsequently, the ship completes intermediate and advanced training exercises as it prepares to deploy.

**Table 2.3**

**DDG-51 Training Requirements by Mission Area**

| Mission Area | Number of Exercises |
|---|---|
| Tactical | |
| Undersea warfare (USW) | 43 |
| Antiair warfare (AAW) | 28 |
| Surface warfare (SUW) | 16 |
| Command, communication, and control (CCC) | 28 |
| Command and control warfare (C2W) | 20 |
| Amphibious warfare (AMW) | 3 |
| Mine warfare (MIW) | 1 |
| Strike warfare (STW) | 3 |
| Nontactical | |
| Intelligence (INT) | 10 |
| Mobility–damage control (MOB-D) | 21 |
| Mobility–engineering (MOB-E) | 38 |
| Mobility–navigation (MOB-N) | 8 |
| Mobility–seamanship (MOB-S) | 23 |
| Noncombat operations (NCO) | 18 |
| Fleet support operations–medical (FSO-M) | 11 |

and whether they were categorized as "must be completed under way" or "could be completed under way or in port" by SURFLANT training officials.

## Exercises That Require Repetition

In some cases, FXP exercises must be repeated to maintain proficiency. Repetition reinforces learned skills and introduces required competencies to new crew members. Figure 2.2 shows the number of training events for various levels of periodicity.

An overall training readiness rating for a mission area is determined by combining the readiness ratings for all exercises in the mission area. More than 50 percent of all required exercises must be repeated at least every three months to maintain M-1 training readiness for these exercises. That is, a ship must repeat more than half of the required exercises every three months to maintain the exercises in an M-1 training readiness rating. About 17 percent of all exercises must be repeated every six months, and 13 percent of exercises must be repeated every 12 months. Thus, about 85 percent of exercises must be repeated within an annual cycle to maintain M-1 training readiness for each exercise.

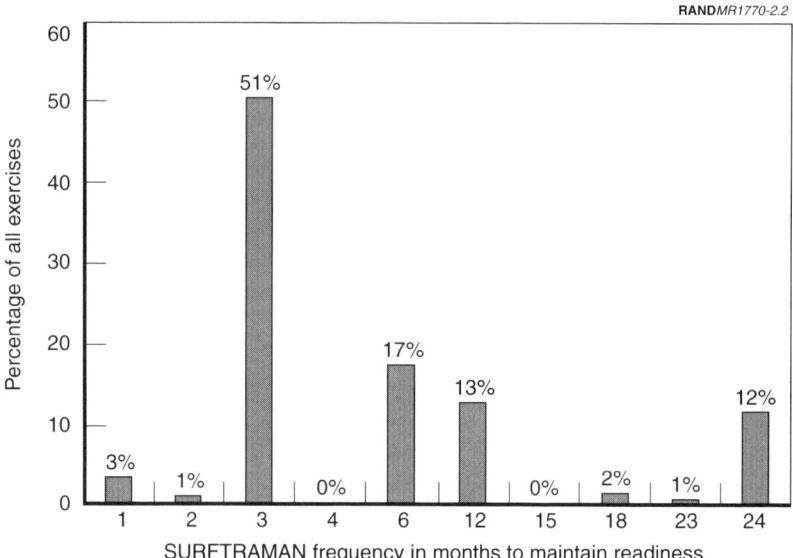

RAND*MR1770-2.2*

NOTE: More than 50 percent of DDG-51 exercises must be repeated at least three months.

**Figure 2.2—Percentage of Exercises That Must
Be Repeated at Each Interval**

## Exercises with Approved Equivalencies

Of the 271 required FXP exercises for the DDG-51 class, only 58 exercises (21 percent) have approved exercise equivalencies for readiness reporting. Equivalencies have been approved for exercises in only six mission areas, all of which are tactical mission areas: USW, AAW, SUW, CCC, C2W, and AMW.

Table 2.4 shows the number of equivalencies in each mission area. There are no approved equivalencies for exercises in nontactical mission areas.

The SURFTRAMAN requires that any exercise for which completion credit is claimed must be performed on the ship's own systems with inputs generated by shore-based or embedded scenario generators or completed through live simulation. This requirement precludes earning "readiness and training" credit by moving such training into shore-based simulators.

### Table 2.4

### Number of Equivalencies by Mission Area for DDG-51 Class

| Mission Area | Number of Exercises | Number of Equivalencies |
|---|---|---|
| Tactical | | |
| USW | 43 | 24 |
| AAW | 28 | 19 |
| SUW | 16 | 6 |
| CCC | 28 | 6 |
| C2W | 20 | 2 |
| AMW | 3 | 1 |
| MIW | 1 | 0 |
| STW | 3 | 0 |
| Nontactical | | |
| INT | 10 | 0 |
| MOB-D | 21 | 0 |
| MOB-E | 38 | 0 |
| MOB-N | 8 | 0 |
| MOB-S | 23 | 0 |
| NCO | 18 | 0 |
| FSO-M | 11 | 0 |

Figure 2.3 shows the number of exercises with equivalencies as a function of the periodicity of the exercises. About one-third of the exercises with the longest duration cycle (24 months) can currently be simulated, but only about one-tenth of those with the shortest duration cycle (three months or less) can be. These findings suggest that simulation has not been applied in areas from which the greatest cost reductions could be obtained. Because one of the benefits of simulation is to be able to repeat training at low cost, it appears that simulation could be most efficiently applied to the exercises that are repeated most often. In principle, such applications would have a higher payback.

## Current Use of Simulators in Surface Force Training

The 1999 SURFTRAMAN authorizes a number of simulators for use in DDG-51 training. Simulators consist of shore-based scenario generators and embedded devices onboard the DDG-51 class ship. Here, we briefly describe each of the simulators, drawing on Navy training manuals and other sources for details about each device.

The **Tactical Advanced Combat Direction and Electronic Warfare (TACDEW) simulator** is a device for task force and team training that can drive up to 22 separate shipboard CIC mockups. At the heart of TACDEW is the Environmental Generation and Control System (EGCS), which generates a synthetic threat environment with up to 2,000 tracks updated at a 1-second rate. The worldwide tactical environment creates ranges from 30,000 feet below sea level to 300,000 feet above sea level and has touch screen user interfaces.

The **Trainer Control Device (TCD)** allows two to eight ships to conduct realistic, multiship antisubmarine warfare (ASW) training simultaneously. Additionally, TCD provides various postscenario tools to enhance the quality of debriefing and further improve the quality of training. The Surface Ship Acoustic Analysis Center (SSAAC) acts as exercise control and transmits the TCD scenarios in accordance with SSAAC's monthly message using TCD modem connections to AN/SQQ-89 onboard-training ships. As a tool that can be used for diverse purposes, TCD provides a valuable example of how technology can meet the Navy's need to maintain warfare proficiencies.

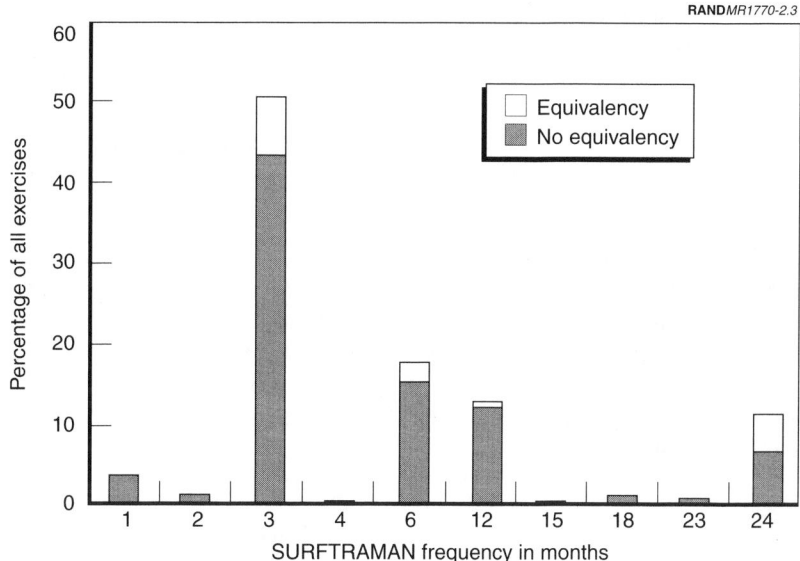

NOTE: Only 10 percent of most frequent exercises can be simulated. This suggests that simulation has not been applied in areas where cost reductions could be obtained.

**Figure 2.3—Percentage of Exercises with Equivalencies Based on Frequency of Required Repetition**

The **Enhanced Naval Wargaming System (ENWGS)** is a computer-based simulator that provides realistic war-gaming in all aspects of naval warfare from the tactical to strategic level, as well as training in decisionmaking for battle group staffs. It supports real-time tactical training of U.S. Navy battle group staffs and students. ENWGS is located at the tactical training groups.

The **AN/USQ-T46(V) Battle Force Tactical Training (BFTT) System** is a highly flexible, interactive tactical combat training system used at the unit, group, and force levels. The BFTT system provides a dynamic, interactive warfighting environment that includes all naval force elements. It supports training of integrated forces or independent ships. The shipboard subsystem training capabilities have been designed around existing onboard or embedded trainer configurations.

The BFTT system wraps around the combat system; both stimulation and simulation of the combat system are transparent to the operators. All controls and displays are in a tactical mode. The combat-system monitoring devices are nonintrusive and have no negative effect on system operation. The BFTT communications subsystem provides a training data link that complies with Distributed Interactive Simulation standards.

The BFTT system is capable of passively monitoring data from the ship's own tactical systems, recording the data for postevent processing, and passively and dynamically replaying the data collected during the simulation to assist in self-assessment. BFTT exercises cannot be conducted at sea. No exercise equivalencies existed for BFTT in calendar year (CY) 2001 because BFTT was just beginning to be installed on ships.

The **20E19 Naval Gunfire Support Mobile Team Trainer** is used to train gun crews by calculating fall of shot without expending live ammunition.

ACTS is an embedded Aegis Weapon System scenario-generating training system that may be used, with or without BFTT, to train CIC personnel. ACTS has greater fidelity than BFTT. Multiship, at-sea training is conducted via ACTS force exercises.

The **Battle Force Electronic Warfare Trainer (BEWT) AN/USQ-T47(V)** is a training support device that stimulates onboard tactical electronic warfare (EW) systems so that afloat operators are able to train for tactical operations using the actual tactical equipment. BEWT enables the onboard training team to exercise individual, team, combat systems team, and force-level operational scenarios with operator keystroke capture capabilities. BEWT provides shipboard proficiency training through stimulation of onboard EW systems including the AN/SLQ-32A(V) and AN/ULQ-16, with future provisions for the AN/WLR-1H(V)7.

The **AN/SQQ-89 Onboard Trainer** is a combat system suite that provides DDG-51 warships with an integrated undersea warfare detection, classification, display, and targeting capability. The system combines and processes all active sonar information and processes and displays all SH-60B Light Airborne Multipurpose System (LAMPS) Mark III sensor data.

The **Cruise Missile Trainer Personal Computer** is a computer program that runs on either the Harpoon Embedded Trainer or any ship-provided laptop. It provides the members of the ship's training team with the ability to conduct Global Command and Control System–Maritime Database Manager training. This system also allows trainers to extract Harpoon engagement planning data from the Harpoon Weapon System. A geographic representation of the training scenario's ground conditions allows dynamic control of the scenario, and a real-time display of the Harpoon missile flyout allows shot evaluation.

A review of the simulators that can be used for exercises with approved equivalencies for the DDG-51 class, as well as those that are used for all other ships, indicated that a number of simulators support the same exercise (see Figure 2.4). For example, in the 1999 SURFTRAMAN, there was one exercise that could be completed on any one of ten different simulators, five exercises that could be com

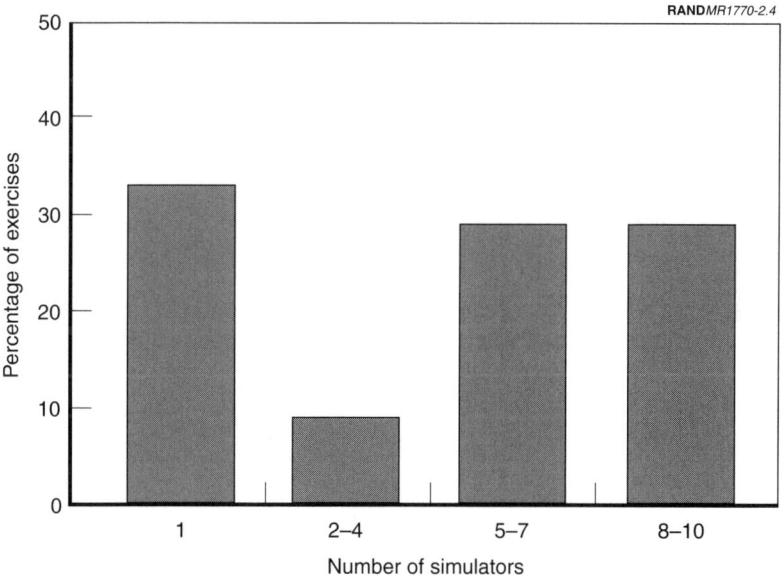

Figure 2.4—Percentage of Exercises That Can Be Simulated as a Function of the Number of Devices Available to Simulate Them

pleted on any of nine simulators, 11 exercises that could be completed on any of eight simulators, and so on.

For each of about 85 percent of exercises that can be simulated, five or more simulators are available. This observation suggests that the Navy training community has chosen to simulate better that which can already be simulated, a pattern of investment that increases costs and may be less likely to produce benefits (in the form of reduced underway time) than would exploring the application of simulation to different exercises.

# Chapter Three

# USE OF SIMULATORS IN OTHER ORGANIZATIONS

Part of our charge, at the outset of this project, was to examine the use of simulation for training in other organizations—both military and civilian—with a view toward understanding both how simulation has been used in those organizations and what issues might arise in efforts to increase the use of simulation for training in the U.S. surface force. Thus, in this chapter, we present results from previous RAND research on the use of simulation in U.S. military training and from interviews with training experts in other military and commercial organizations. These findings enabled us to specify opportunities for increasing the use of simulation for training and to articulate issues that must be considered in such efforts.

## USE OF SIMULATORS IN THE AVIATION COMMUNITY

Simulators are used extensively for training in the aviation community, but they cannot simply substitute for flight time. The aviation community's experience indicates that the value of any training method, whether live or simulated, depends, among other things, on task characteristics and complexity.

For instance, in a recently published report on training for fighter strike missions (Schank et al., 2002), we observed that a relatively small proportion of training exercises is conducted through simulators. This was true for the U.S. Navy and U.S. Marine Corps F-18s, as well as for U.S. Air Force, Royal Air Force, and French Air Force fighter training. This pattern is, in part, a consequence of the limited availability of appropriate simulators and the poor fidelity of those that are available. To the extent that simulators are used, they are used by those in the training pipeline before the pilots report to operational units.

For maritime patrol aircraft (MPA), however, about 50 percent of both basic flight and mission training exercises are completed on simulators. This finding is based on observations in the U.S. Navy, as well as in the Royal Air Force and the French Air Force. Training experts in these areas report that keeping mission-oriented simulators up to date is problematic. In commercial aviation, nearly all training is completed on simulators.

Training experts with aviation experience also observed that, for some kinds of training, simulation was seen as useful for practice, but verification of proficiency remained a live event. The French MPA community, for instance, shifted from 15 actual flights to ensure aircrew qualifications to a program of 10 to 12 simulator events, followed by two live flights, the second of which is the examination. In general, the value of simulators for rehearsal is greater for junior personnel than for senior personnel. Simulators can, however, also be useful for experienced pilots, who use them to show currency in particular areas without having to fly the event.

In addition to drawing attention to the distinction between rehearsal and qualification, the French MPA example illustrates another important point: It may be useful to structure some kinds of training as a combination of simulated and live events. Simulator-only training may be best for high-risk events or events that are difficult to conduct live, e.g., assessing the safety-of-flight events.

Simulation was regarded as effective in such areas as introducing and setting up equipment and "switchology," but live training was viewed as superior to simulation for learning perceptual motor skills. Finally, it was noted that simulators can become predictable and can teach some inappropriate skills, resulting in a false sense of accomplishment.

These observations offer useful insights, but, despite extensive use of simulation within the aviation community, there is, as yet, no clear evidence about what constitutes the optimal combination of simulation and live training for specific training exercises. Moreover, it is important to note that, although the specific practices adopted in the aviation community suggest avenues for increasing the use of simulation, the unique training requirements of the surface force would need to be taken into account in applying these insights in the context of surface force training.

## USE OF SIMULATORS FOR TRAINING IN THE CANADIAN NAVY

The Canadian Navy is a small force, with 13,000 personnel and 30 ships, including four *Iroquois*-class destroyers and 12 *Halifax*-class frigates. The fleet is divided into Atlantic and Pacific forces. The Maritime Forces Atlantic consist of seven *Halifax*-class patrol frigates, two *Iroquois*-class Destroyers, six *Kingston*-class coastal defense vessels, one operational supply ship, and one sail-training yacht. The Maritime Forces Pacific consist of five *Halifax*-class frigates, two *Iroquois*-class destroyers, six *Kingston*-class coastal defense vessels, one supply ship, and one sail training yacht. Three of four *Upholder*-class diesel submarines have been accepted as the *Victoria* class and will come into operational service on both coasts over the next three years.

Canada provides ships to NATO, U.S. Navy carrier battle groups (CVBGs), and international operations. Canadian vessels are the only foreign vessels that the U.S. Navy has allowed to fully integrate into its operational groups.

Canadian ships are maintained in one of two readiness states: *high*, meaning that the ship must be ready to deploy in ten days, and *standard*, meaning that the ship must be ready to deploy in 30 days. Ships in a high state of readiness must maintain currency in a minimum of 90 percent of training events, and ships in a standard state of readiness must maintain currency in a minimum of 75 percent of training events. The ready ship alternates between coasts. Ships steam an average of 90 to 120 days a year, depending on their readiness status. After September 11, 2001, the East Coast ship was the ready ship and deployed in support of Operation Enduring Freedom.

Readiness requirements are detailed in the *Maritime Command Combat Readiness Requirements* (Chief of the Maritime Staff, 2001). These requirements are structured in four levels:

Level "A" serials are basic training events designed to enhance individual operator skills; they may be conducted during training at pierside, in tactical and procedural trainers, or at sea.

Level "B" serials are intermediate training events designed to enhance the readiness of specific teams (e.g., above-water warfare team, chief information officer, etc.); they may also be conducted at pierside, in trainers, or at sea.

Level "C" serials are advanced training events involving complete teams (e.g., the starboard watch). Although some Level "C" combat readiness requirements (CRRs) may be completed at pierside or in trainers, they should be conducted at sea, where feasible.

Level "D" serials are advanced training events that focus on the assessment of groups of ships (normally the task group). These are considered task group serials and are not normally reported by individual units. Completion of these CRRs is important for achieving and maintaining readiness levels. These CRRs are conducted at sea.

Underway training is required for only the most advanced exercises—the multiship exercises in Level D. Examples of such exercises include missile firings and ASW training and torpedo firing at the Atlantic Undersea Test and Evaluation Center range in the Bahamas and complementary ranges on the West Coast. For all other levels, the use of simulation for training is both permitted and encouraged.

The Canadian Navy's reliance on simulators for training may be related to the limits on the number of underway days available for training. Our communications with Canadian Navy officials indicated that their fuel budget dictates how much steaming they can do. Independent steaming time (for training) is minimized for budgetary reasons, and training and team qualifications are completed through simulation, pierside exercises, or some combination of the two. The use of simulators fills that gap that the constraints on underway training time cause. When ships in the Canadian Navy get under way, they do so to participate in a specific major exercise or operation, both domestic and international.

Some simulators are at fixed sites, with personnel traveling to them for training. For instance, at the Naval Operations School in Halifax, Nova Scotia, the Canadian Navy has a team trainer for the *Halifax* class of ships; this trainer is similar to the U.S. Navy's Aegis trainer at Morristown, NJ. The Canadian team trainer is referred to as the "13th ship" and has a replication of the "box room" (operations center).

There is a similar trainer for the *Iroquois* class (the "5th ship") on the West Coast. This trainer is for various operators on the ship and includes all warfighting capabilities up to actually firing weapons. It has both simulation and stimulation capabilities and is in almost constant use. Scenarios involving EW, ASW, and SUW are generated for the watch teams, and a fixed amount of time is allotted for the watch teams to sequence through the scenarios.

The trainer for the *Iroquois*-class ship was built as part of the procurement contract for the ships, a practice the Royal Navy has also adopted. This practice permits integration of simulation in the design of the ship and in plans for training. After-the-fact planning decreases the likelihood that funds will be available to support the design of appropriate simulators and increases the cost of designing and building them, as well as the cost of integrating them in training exercises.

The Canadian Navy uses multiple simulators for diverse purposes. As noted previously, some of these simulators are located at sites on the East and West Coasts of the country, with personnel traveling to them for training. Various ships also have onboard trainers. A sea training group rides with the ship to evaluate the effectiveness of onboard training. The group remains onboard for about two weeks, assessing combat readiness in addition to conducting exercises, such as lighting flares to simulate fires to see how the crew reacts. The whole ship is evaluated to determine whether it is ready to assume its readiness status.

An engineering training facility, which houses an Integrated Machinery Control System provided as part of the *Halifax*-class contract, is used to familiarize engineering personnel with the system before they go to sea. In addition, there is a maintenance procedures trainer (Lockheed Martin) for the *Halifax* class.

The Canadian Navy also has a navigation and bridge simulator that can simulate the ship-handling characteristics of all Canadian ships, which, according to Canadian Embassy officials, is used "around the clock." The bridge simulator provides a nearly 360-degree field of view. This facility is also used to train radar navigation teams to work with radar displays without the aid of visual references; merchant marines also use the facility to complete their own training requirement.

## USE OF SIMULATORS FOR TRAINING IN THE ROYAL NAVY

Royal Navy training is similar to that of the U.S. Navy, in that it is incremental, progressing from individual training to subteam training to full-team training to full-team training with the principal warfare officer and, finally, to full-team training with the principal warfare officer and the commanding officer. Thus, observations regarding their use of simulators for training, including both advantages and disadvantages, should be relevant to training in the U.S. surface force.

In the Royal Navy, underway time is reserved for training in primary warfare mission areas. Underway training includes simulated events, such as Harpoon console training. This training can be combined or made more challenging by using helicopters to transmit pictures of targets, with simulators providing experience with situations in which an enemy ship returns fire.

In other areas, simulation is used to reduce costs and risks. For instance, the Royal Navy has a navigation and ship-handling simulator, as well as a CIC mockup for every class of ship. For Type 22 and 23 ships, the Royal Navy uses these trainers for new officers and also brings CIC teams from existing ships to HMS *Dryad*, one of the Royal Navy's largest training facilities, to run tactical exercises as a prerequisite for operational sea training at Devonport. Operational sea training is used as a postrepair or predeployment training period and provides the Royal Navy readiness certification process. The combat system suites on the Type 22, 23, and 45 ships all have a simulation or training mode.

The Royal Navy has seen a steady increase in underway training for command and control events, but there is less programmed time for engineering training. Simulator training for the engineering teams is usually carried out during the latter weeks of a programmed docking period for major repair and update work. Engineering simulators are in the Portsmouth shore establishments and only the Type 23 simulator is considered to be realistic. Training for "general quarters" action damage is also carried out in simulators in the Portsmouth area, these being of high quality. Unfortunately, the geographic location means that ship's teams based in Plymouth and Scotland use these simulators less than desired. For engineering teams, underway training is the more generally accepted method of raising operational competence, taking place during operational sea training (external training help) or during sea transit time (self-training). Royal Navy training officials believe that the availability of a Type 23 engineering trainer in each port could help to reduce this demand for underway time. This is consistent with our observations regarding simulation in the aviation community. That is, to obtain the maximum benefit from simulation, enough simulators must be available to provide adequate training time for all relevant personnel.

## USE OF SIMULATORS FOR TRAINING IN COMMERCIAL SHIPPING

During our research, we visited the Maritime Institute of Technology and Graduate Studies (MITAGS). There, we learned that private industry is overhauling the qualifications for mariners. Private industry standards are established through the International Convention on Standards of Training, Certification, and Watchkeeping (STCW) for Seafarers, and the United States is a party to this convention. The most recent convention (1995) requires a hands-on demonstration of skills and ability to serve aboard seagoing vessels. The convention also formalized the documentation of a mariner's ability to perform key tasks.

The U.S. Navy is developing training programs that will allow naval personnel to earn a qualification that conforms to the merchant marine standard. The anticipated changes will capitalize on the efforts of the commercial shipping industry to establish standards and required qualifications of watch standers. The changes will also align U.S. Navy standards for accreditation of simulators with those of private industry and help to achieve the Chief of Naval Education and Training's goal of accreditation of personnel to civilian standards. This certification will ease the transition of naval personnel to civilian jobs and, it is hoped, will also foster transitions between the private sector and the military by easing the lateral transfer of civilian personnel into the military and of military personnel to private industry.

The STCW convention establishes minimum requirements for personnel in certain areas of responsibility, including the following:

- master and deck department

- engine department
- radiocommunication and radio personnel
- personnel on certain types of ships.

The STCW also establishes more-general mandatory minimum requirements for

- familiarization, basic safety training, and instruction for all seafarers
- certificates of proficiency in survival craft, rescue boats, and fast rescue boats
- training in advanced firefighting
- medical first aid and medical care.

In the Navy, under Task Force Excel, curriculum changes will conform to international standards of training. The STCW international standard requires personnel to be qualified in tasks required for specific watch stations and permits these qualifying tasks to be performed on appropriate simulators.

In regard to the benefit of simulation in training, MITAGS representatives said that the simulators allow everyone to be trained through a core set of drills in a timely manner. In addition, simulators allow personnel to experience the casualty control exercises and procedures not normally done at sea, thereby preventing equipment damage and personal injury.

MITAGS also operates a ship-handling simulator; the representatives we interviewed said that these simulators enable officers to maintain ship-handling proficiency and to develop expertise in special operations. Training for emergencies, which is normally too dangerous to be completed aboard ship, can also be done in a simulator. Using simulators to practice planned responses to emergency ship-handling situations also helps crew members prepare for handling casualties, which cannot be rehearsed easily under way. The advantage of ship-handling simulators is that they allow several crew members to cycle through the training, make mistakes in a forgiving environment, and learn from their mistakes.

The experience of other organizations, both military and civilian, indicates that the use of simulation for training in the U.S. Navy surface force could be substantially increased. Further, we were able to draw a number of lessons from our discussions with training officials in these organizations about the kinds of policies and training practices that would be needed to support effective use of simulation in the surface force.

First, standards that specify levels of proficiency, without regard to the method of training, are needed. Without such standards, it is impossible to assess the relative merits of live and simulated training.

Second, investments in simulation must be adequate to ensure that all relevant personnel can complete the needed training in a reasonable time. Inadequate simulation resources will reduce the efficiency of the training even if the simulators provide high-quality training. Resources can be stretched by planning for simulation while designing ships and developing training exercises and by positioning simulators at a few sites and moving people to the simulators, as is done in the Canadian Navy and the Royal Navy.

Third, the use of live training and simulation must be balanced to obtain optimum effects. Achieving such a balance might involve, for instance, defining the respective roles of simulation and live training within a given exercise, determining how simulation might be used in a set of sequentially or hierarchically linked exercises, or specifying parameters for the use of simulation based on current levels of proficiency.

Finally, cultural bias with respect to simulation must be overcome. In military aviation, simulator hours are accorded little respect, and the same is generally true in the surface force. To increase the perceived value of simulation, its role in attaining proficiency and readiness must be firmly established at the highest levels of the organization.

# Chapter Four

# U.S. NAVY VIEWS ON SIMULATION AND ROADBLOCKS TO SIMULATION

In this chapter, we report the results of our discussions with Navy officers regarding the possibility of increasing simulation in training for the Navy's surface force. We also describe current obstacles to increasing use of simulators and discuss recent progress in that area.

## FLEET OPINIONS ON SIMULATION

During our investigation, we discussed possibilities for improving current simulation practices and for increasing the use of simulation as a replacement for live training with surface force training officials. The representatives of the Commander, Naval Surface Force, Atlantic (COMNAVSURFLANT) and the Commander, Second Fleet (COMSECONDFLT) were open to these prospects; they indicated, however, that much of the "fat" has already been taken out of operational training. COMSECONDFLT representatives added that further reductions in underway training may risk professional competence but did not specify metrics that could be used to assess professional competence—regardless of where or how training was conducted.

Fleet training authorities indicated that completion of underway exercises is regarded as paramount in the development of readiness; they reported that most members of the surface community believe that reducing underway time will reduce readiness. For instance, in numerous articles, COMLANTFLT has observed that simulation can be used to augment training completed under way but will not replace it.

Fleet authorities also reported that scheduling in-port training is challenging because of equipment maintenance requirements and personnel availability. In addition, scheduling authorities must deconflict major events to ensure participation in BFTT in-port exercises. For example, the scheduling authority had to schedule an in-port exercise four different times before the event could be conducted by three ships in port. If more training is moved into port, questions of how to handle scheduled maintenance and other activities would have to be addressed.

Despite having acknowledged the potential value of simulation, COMSECONDFLT and COMNAVSURFLANT disagreed about how it might best be used to reduce underway training. The COMNAVSURFLANT representatives indicated that much of the underway time for intermediate and advanced training was used for training deploying staff, which could be completed in port. COMSECONDFLT representatives were protective of the underway time for ships in the intermediate and advanced phases of training, suggesting that efforts to reduce underway time during the IDTC should focus on ships in the basic training phase.

COMSECONDFLT personnel did not seem to emphasize simulation. Although Naval Surface Fire Support (NSFS) training is moving toward simulation through Virtual At-Sea Training (VAST), which is done under way, and BFTT is being tested in efforts to replace underway time in the intermediate and advanced phases, fleet representatives indicated to us that, for advanced and intermediate training, ships need to get under way. In their view, in-port battle group exercises enhance underway training, but they do not replace underway training requirements.

## DIFFICULTY IN ASSESSING READINESS

An important barrier to increasing the use of simulation is the difficulty of assessing a ship's readiness to deploy. Historically, this evaluation has been largely subjective. In the basic phase, ships must complete 80 percent of their required FXP exercises and must satisfactorily complete the FEP. In the FEP, the ship's training teams must demonstrate the ability to self-train. Evaluation of this ability is based on the judgment of the ship's ISIC, assisted by ATG personnel during the FEP. In the intermediate phase, readiness is also evaluated by the ISIC. In this phase, as well as in the advanced phases of training, evaluations are more subjective than in basic training.

Many of the senior officers now in charge of evaluating intermediate and advanced training believe that the expertise required to evaluate its success cannot be transferred to paper. That is, they believe these complex judgment processes cannot be standardized so that they can be taught to less-experienced personnel. This belief is, however, more strongly held with regard to some kinds of training than others. In such mission areas as ASW and SUW, for instance, the success of training is not hard to judge, but assessing the efficacy of training is more difficult in such areas as damage control and engineering casualty control.

Under the *sea combat commander* concept, a DDG-51's ISIC specifies a ship's readiness and training requirements. The ISIC will deploy with the CVBG, will be the DDG-51's warfare commander, and will determine the unit's readiness to perform its primary warfare missions. The ISIC assesses the ship's ability to support and perform these missions in the Composite Training Unit Exercise, and the numbered fleet commander is responsible for certifying the CVBG in 13 mission areas upon completion of the Joint Task Force Exercise.

In corresponding with a Navy surface training official, we were told that no appropriate metric exists that compares the value of live against that of simulated training, and some officers doubt the feasibility of developing such standards. This belief may not be universal, but it may reflect an underlying reservation that is widely held and likely to affect acceptance of training through simulation.

One factor complicating assessments of readiness is the misalignment of personnel rotation schedules and deployment schedules. That is, because training schedules and deployment schedules are not synchronized, a ship may complete training exercises with one crew, a portion of which may then rotate off the ship and deploy with another crew, some of whom have not had the benefit of all phases of training. This pattern undermines efforts to obtain valid assessments of readiness, perhaps inflating judgments of readiness that are made based on the performance of the more experienced, but now departed, personnel.

In fact, we learned from prior commanding officers that personnel rotations are sometimes held up until key training events, such as the FEP, are completed. After passing the FEP or other key event, personnel are then allowed to rotate off the ship, to be replaced by personnel who may be less skilled than the people they replaced. Although alignment of personnel rotations with a ship's training workups and deployment schedule is a stated goal of the fleet training strategy, current patterns of personnel rotation undermine accomplishment of this goal. The personnel that the ship trains and deploys with need to be aligned to maximize not only the continuity of training and readiness but also the validity of assessments of readiness.

## ROADBLOCKS TO INCREASED USE OF SIMULATION

Many factors can interfere with improved or increased use of simulation. To achieve the possible benefits of increased simulation, the effects of these factors must be taken into account in developing new simulators and new uses for simulators. Here, we identify several such factors and describe the issues they raise.

### Training Policies That Preclude Earning Credit for Shore-Based Training

As previously noted, any exercise for which completion credit is claimed must be performed on the ship's own systems. This requirement precludes earning "readiness and training" credit by moving such training into shore-based simulators. To take full advantage of the capabilities of current simulation technology, this requirement would need to be modified. In terms of naval policy, this may be the most important roadblock to increasing the use of simulation.

### Cultural Bias

Navy officers acknowledge a bias toward completing training exercises under way. For instance, a fleet training representative told us that we would "find no officers with stars on their collars [who would] say that they would trade a day of under way training for a day of training in port." This bias is reflected in the patterns we observed in our analysis of the locations in which training exercises were completed, which the next chapter will describe. Our findings indicate that even exercises with approved equivalencies or that, as indicated by fleet training representatives, could be completed in port were actually completed under way in a substantial majority of cases. A change in Navy culture may be required to achieve a higher level of in-port training involving simulated exercises.

## Cost

Simulators are expensive, and, depending on the goals of training and on how simulators are incorporated into current training policies and practices, it is conceivable that their use could increase, rather than decrease, overall training costs. Such an outcome could come about in one of two ways.

The first possibility has to do with the extent to which training exercises are completed while ships are deployed. Our analysis indicates that, of all exercises DDG-51 class ships completed during the period under study, deployed ships completed 31 percent and completed 99 percent of these while under way. These observations raise questions about the economic benefit of simulating exercises that are typically completed while ships are under way. In particular, they suggest that careful consideration should be given to the use of simulation as preparation and practice for exercises that could then be completed more efficiently when ships are deployed and hence under way most of the time.

The second possibility has to do with increasing the use of simulation for training on nondeployed ships. If the present practice of spending an average of 28 days per quarter under way for training while nondeployed is maintained, the cost of simulators would be added to the cost of underway time, thus increasing the cost of training on these ships. Thus, achieving economic benefits by increasing the use of simulation for training on nondeployed ships requires reducing underway days. The analysis we will describe in the next section indicates that there are numerous opportunities to decrease the proportion of underway training on nondeployed ships, either by completing more exercises with equivalencies in port or by completing more of the exercises in port that can be completed either in port or under way.

Of course, if it can be shown that simulation provides an advantage over live training in terms of proficiency and if increased proficiency is deemed a policy goal, it may be that the added cost of simulation is justified. In either case, the cost and value of simulation can only be determined in relation to specific training goals, policies, and practices.

## Absence of Standards for Assessing Readiness

FXP exercises are reported in the Training Readiness Management System (TRMS), which is used to compute the ship's training readiness in the SORTS reporting. This record is used to assess a ship's training readiness at the basic training phase. As we have indicated previously, however, the particular mix of exercises completed during intermediate and advanced training is determined by the ship's ISIC, the deploying battle group commander, and the numbered fleet commander, who judge a ship's readiness to deploy. COMSECOND-FLT representatives told us that ships are trained and assessed to Navy Mission Essential Task Lists requirements but that SORTS does not adequately capture these requirements. Thus, there is a disconnect between measures of training readiness in SORTS and other measures of training readiness that ships must meet before deployment.

This disconnect is important in itself but also has important implications for determining which exercises might be simulated and for assessing the value of simulation. The Navy needs better measures of readiness for its own sake, as well as for the sake of having criteria for selecting exercises for simulation, designing simulators, and assessing proficiency levels attained through simulation.

## Interference with Other Activities

The amount of time a ship spends in port could be adjusted to correspond to new policies regarding the location of training, but there are many activities that limit the amount of time available for in-port training. These activities include maintaining and repairing equipment and systems and onloading stores, as well as meeting such competing demands on personnel as medical and dental appointments, family needs, schools, physical fitness training, and other training.

Reducing underway time to devote more time to in-port training on simulators would also interfere with activities that require that the ship be under way. For instance, ships need to be under way to support other fleet training requirements, such as serving as an opposition force ship in the training of deployers and supporting deck-landing qualifications for helicopters. Further, ships must be available for community-relations port visits and guest-of-the-Navy cruises, and these activities limit the time available for in-port training.

## Competition Between Units for Use of Simulators

The demands on a ship's crew limit their availability to perform training ashore. Other units ashore at the same time will be competing for simulator availability to complete their training requirements. Thus, as we discussed in relation to the aviation community's experience with simulation, ensuring the effective use of simulation will require that enough simulators be available to train all relevant personnel when they are available for training. Ensuring this availability will, of course, affect the costs associated with increasing the use of simulation in training.

## PROGRESS IN THE USE OF SIMULATION FOR SURFACE FORCE TRAINING

Despite the numerous obstacles to increasing the use of simulation in surface force training, substantial efforts to achieve this goal have been launched. For instance, COMSECONDFLT was tasked with making in-port integrated warfare training a reality and a standard requirement for battle group interdeployment training.

The organization developed a concept of operations delineating the responsibilities of and procedures for ships, staffs, and assisting commands to plan, develop, generate scenarios, transmit, control, and reconstruct and evaluate Battle Group In-Port Exercises (BGIE). A trial phase was conducted from January 2002 through spring 2003. The plan is that, eventually, battle groups will conduct the BGIE as part of their IDTC training.

The BGIE design consists of a series of in-port, onboard training exercises that provide multiwarfare training at various levels, from the individual unit through the battle or amphibious ready group commander, using existing onboard trainers and BFTT and shore-based systems. The five BGIE levels have been designed to help operators, ships, and battle or amphibious ready groups meet their individual training objectives and requirements. Phases 1 and 2 are completed during basic training; Phases 3 and 4 are completed during intermediate training; and Phase 5 occurs during advanced training. All scenarios use the BFTT–TCD–multiunit tactical training system architecture.

The warfare area varies for each exercise, to allow training teams to focus onboard training on specific areas. Surface ships must conduct three exercises in their primary and secondary warfare areas during Phase 1. The exercises in Phase 2 are basic training exercises that employ units in a multiship environment.

Phase 3 exercises provide battle group warfare commanders an opportunity to train their teams in all applicable warfare areas. The exercises consist of scenarios battle group commanders themselves have developed to test battle group operational tasks. Phase 4 consists of intermediate exercises designed to provide the fleet commander an opportunity to evaluate the tactical proficiency of the battle group.

Phase 5 is an advanced exercise intended to improve proficiency in specific areas that the numbered fleet commander designates, giving that commander an opportunity to refine the battle group's tactical proficiency in specific warfare areas prior to deployment.

Overall, a significant amount of time will be invested in conducting BGIE. It has not yet been determined whether these exercises will be conducted in addition to underway exercises or whether they will replace underway training requirements. When we asked about this issue, COMSECONDFLT representatives were noncommittal, indicating that BGIE is in the testing phase and that no decisions regarding its relation to current training requirements had been made. However, it appears that the Navy has increased its emphasis on in-port training using BGIE, because ships preparing for deployment are required to complete these exercises.

Two additional examples of the move toward greater use of simulation are the use of the Conning Officer Virtual Environment (COVE) and VAST. We next describe each of these simulators.

## Conning Officer Virtual Environment

COVE is a device for learning and practicing ship-handling skills that depend on "Seaman's Eye"—the ability to interpret wind, current, ship's speed, and a combination of other visual factors—which includes understanding ship dynamics, interpreting perceptual cues, and other information available to a conning officer on the bridge and, based on this understanding, applying rules of thumb for responding to situations that arise while maneuvering.

COVE can provide many benefits. In addition to providing opportunities to practice Seaman's Eye, COVE can be deployed in the schoolhouse as well as in the fleet and eliminates the need for other crew members to be present while practicing.

## Virtual At-Sea Training

The Atlantic Fleet has begun testing the VAST system, a portable simulation device that will let sailors train—in port or at sea—with advanced weapon systems. The trainer will allow crews to train and rehearse with real or simulated ordnance. The system simulates terrain by superimposing targets over an area of water. The images are displayed in three dimensions on one computer, with a second computer providing a two-dimensional map and other data. Trainers can use a simulated unmanned aerial vehicle to provide reconnaissance of targets.

Using VAST, ship or shore spotters identify and communicate the locations of the virtual targets to the ship, which then aims and fires its weapons. When a ship fires into this array, acoustic sensors detect the impact of the round and transmit the resulting data to the system control computer aboard the ship. The buoys, known as the Integrated Maritime Portable Acoustic Scoring and Simulator (IMPASS), comprise the Global Positioning System equipment, a hydrophone, processing circuitry, and a battery power supply. The computer then calculates the precise point of impact within the IMPASS array and passes the results to the system in real time.

This technology could provide a means of rehearsing "in" the areas where it is believed that attacks are likely. It can also provide more flexibility for training in local operational areas as required, instead of having to sail great distances to a range, such as Vieques, to complete this qualification. Fleet representatives indicated that this technology is not intended to replace live training events. Instead, VAST and IMPASS would be used to assist crews in preparing for live-fire exercises and to help them achieve an overall higher level of expertise.

# Chapter Five

# ANALYSIS OF TRAINING LOCATION

In this chapter, we analyze the relationship between the location in which exercises *could* be completed and the location in which they *were* completed. In terms of where they could be completed, there are three classes of exercises:

- exercises with equivalencies

- exercises with no equivalencies that can be completed under way or in port

- exercises with no equivalencies that can only be completed under way.

Our analysis specifies the location in which each of these three classes of exercises was completed—under way or in port—during the period under study.

## WHERE EXERCISES WERE COMPLETED

To determine where training exercises were completed, we created a database that linked the ship's employment data (dates when DDG-51 class ships were at sea or in port) to the dates of reported exercise completion. By connecting the location of ships to the reported completion of exercises, we were able to determine whether any given exercise was completed when the ship was under way or in port.

As expected, our preliminary analysis indicated that, when ships are deployed, 99 percent of required exercises are completed under way. Therefore, because we were interested in identifying opportunities to increase the use of simulation for training and, particularly, in possibilities for using simulation to reduce underway training time, our analysis focused on nondeployed ships. It is these exercises that offer the greatest potential for reducing costs by completing more training exercises while in port.

We focus here on the distinction between all exercises and exercises with approved equivalencies. We anticipated that, compared to the set of all exercises, a higher proportion of the exercises with approved equivalencies would be completed in port.

## Data Sources

To complete this analysis, we needed four kinds of information: the location of ships on specific dates during the period under study; exercises that were completed and the dates on which they were completed; whether or not each of the completed exercises had an approved equivalency; and whether, in the views of COMSURFLANT training officials, exercises with no equivalencies could be completed under way or in port or could only be completed under way.

**Location of Ships.** To obtain information about the location of ships at any given time, we used the COMLANTFLT employment schedules, which provided information about the activities for which a ship was employed on a given date. The Third Fleet employment terms allowed us to classify each employment term as under way, in port, or in port steaming (COMTHIRDFLT, 1998).

**Exercise Completion Reports by Date.** To obtain information about exercises that were completed and when they were completed, we used the type commanders' TRMS. Training readiness C/M-ratings reported by SORTS are determined by training reports (TRNGREPs) submitted by ships and compiled in TRMS.[1] TRMS is updated when a ship submits TRNGREPs. TRMS uses the TRNGREP data to convert exercise completions into exercise M-ratings and to calculate mission area training readiness M-ratings based on the number of exercises completed in each mission area. Ships submit TRNGREPs

---

[1] *C ratings* describe a unit's overall readiness status; *M-ratings* relate to readiness in a specific mission area.

monthly and upon completion of at-sea training periods, significant exercises, and inspections.

**Classification of Exercises Based on Approved Equivalencies.** To determine whether an equivalency had been specified for each of the completed exercises, we consulted the SURFTRAMAN.

**Classification of Exercises Based on Judgments of Where Exercises Could Be Completed.** We asked COMNAVSURFLANT N7 Training and Readiness personnel, who are responsible for planning and overseeing surface force training for the Atlantic Fleet, to tell us which of the exercises for which there were no equivalencies could be completed in port and which could only be completed under way.

These officers indicated that 65 of the required exercises for the DDG-51 specified in the SURFTRAMAN could only be completed under way.[2] These "underway only" exercises consisted primarily of live firing, navigation, seamanship, and engineering exercises.

## Data Collection

As indicated previously, we limited our analysis to a single ship class, the DDG-51. We selected this class to meet two criteria: the feasibility of obtaining the data needed to conduct the analysis and the likelihood that such an analysis would yield results that might be useful not only for designing training for ships in that class but also for the development of plans for training on other kinds of vessels.

The DDG-51 is a new ship and has a simulation capability, and a relatively large number of ships were available, providing both a platform with relevant features and a sample size adequate for our research. Thus, the results of our research have both validity and generalizability. Based on a large number of exercises, the quantitative evidence we present is representative of the real training experience of these ships over time. And, because of this ship's simulation capability and its status as a new ship, our results should be applicable to and useful in planning for training on ships that will be in active service for years to come.

---

[2]Appendix A of this report shows the exercises that must be completed under way as opposed to those that could be completed in port.

## Construction of the Database

Using the data we collected, we created a database containing

- the name of each ship
- the name, number, and mission area of each exercise
- an indication of whether the exercise had an approved equivalency
- an indication of whether the exercise required repetition and, if so, at what interval (e.g., three months, six months)
- where, in the judgment of Navy training officials, exercises without equivalencies could be completed (i.e., in port or under way)
- the dates on which exercises were completed
- the location of the ship (under way or in port) on each day throughout the period under study.[3]

This database enabled us to determine where an exercise was completed (in port or under way) based on the date the ship reported completing the exercise. Further, it enabled us to distinguish the location of different types of exercises, including those

- in different mission areas
- that required repetition
- that had approved equivalencies
- with no equivalencies that, according to Navy training officials, could only be completed under way
- with no equivalencies that, according to Navy training officials, could be completed either under way or in port.

_____

[3]In some cases, a ship could have been under way for part of the day prior to mooring to pier. In these cases, the data were scrubbed, using our best judgment, to determine where specific exercises were likely to have been completed. In making this determination, we erred on the side of exercises having been completed under way.

## ANALYSIS OF EXERCISE COMPLETION

As we have indicated, our analysis focuses on exercises completed by nondeployed ships because our primary interest is in where exercises of various kinds were completed. When ships are deployed, training exercises are almost always completed under way.

Nonetheless, to provide a comprehensive view of the distribution of training exercises—across ships and locations—we will first describe the pattern of exercise completion across all ships; then by deployed ships; and, finally, by nondeployed ships.

### All Ships

Our database contained more than 12,000 FXP exercises completed during CY 2001 by 18 Atlantic Fleet–based DDG-51s.

There was considerable variation in the number of exercises completed across the ships of the class. USS *McFaul*, USS *Arleigh Burke*, and USS *Ramage* each completed more than 1,000 exercises during the year, while USS *Donald Cook*, USS *Carney*, USS *Oscar Austin*, and USS *Roosevelt* each reported fewer than 600 completed exercises.

This variation can be attributed to the ships being in different stages of readiness during the period under study. Some ships were in maintenance periods and hence reported a smaller number of exercise completions; some were in basic training; and some were in intermediate or advanced training. In addition, eight ships were deployed during some part of CY 2001.

USS *Bulkeley* was commissioned in December 2001 and thus contributed few observations to our database. USS *Porter* reported completing only 116 exercises. Although it was deployed for five months, this number is unexpectedly low given its long stay in port.

There was also variation in the numbers of exercises completed in port among ships of the class. These exercise completions ranged from a high of 256 (USS *Mahan*) to a low of approximately 30 exercises completed in port for USS *McFaul*, USS *Gonzalez*, and USS *Mitscher*.

Across all ships, 84 percent (10,195) of all exercises were completed under way, with 16 percent (1,962) of FXP exercises being completed in port.

## Deployed Ships

Eight ships were deployed during CY 2001. Four ships completed a full six-month deployment, and four were on a six-month deployment for part of the year.

Of the more than 12,000 exercises reported as completed by DDG-51s for CY 2001, 3,765 (31 percent) of all exercises were completed by the eight ships that were deployed for part or all of the year during CY 2001.

Of all exercises completed by deployed ships, 99 percent were reported as having been completed under way.

## Nondeployed Ships

Table 5.1 summarizes the data for nondeployed ships. Based on the records of exercise completion we used, 8,250 exercises were completed by nondeployed ships; of these, 6,356 exercises (77 percent) were reported as having been completed under way, and 1,894 (23 percent) were reported as having been completed in port. (See Appendix B for more details regarding the proportion of exercises with equivalencies that were completed in port, organized by mission area.)

As was the case with all ships, there was considerable variation within the group of nondeployed ships in the proportion of exercises reported as having been completed in port. USS *Mahan* completed 38 percent of the exercises it reported in port, the highest percentage among nondeployed ships. USS *The Sullivans* completed only 10 percent of its exercises in port.

**Table 5.1**

**Where Exercises with Equivalencies Were Completed**

| Type of Exercise | Location | |
| --- | --- | --- |
| | In Port (%) | Under way (%) |
| All exercises[a] | 77 | 23 |
| Exercises with equivalencies | 80 | 20 |

[a]All figures are for nondeployed ships only. For deployed ships, 99 percent of all exercises were completed under way.

For exercises with equivalencies, the proportion of exercises reported as having been completed under way or in port is nearly identical to the proportion of all exercises completed under way or in port. **Of exercises with equivalencies, all of which could be completed in port, 80 percent were reported as having been completed under way. Only 20 percent were reported as having been completed in port.**

Thus, contrary to what might be expected, the proportion of exercises with equivalencies completed under way by nondeployed ships was about the same as the proportion of all exercises completed under way. **The existence of an approved equivalency did not affect the likelihood that an exercise would be completed under way; exercises with and without equivalencies were equally likely to be completed under way.**

When we examined the location in which exercises with equivalencies were completed on the basis of mission area, we found that the proportion of exercises with equivalencies completed in port did not differ greatly across mission areas. Most of the exercises with equivalencies are in three mission areas (see Table 5.2). Thus, mission area does not appear to determine whether exercises are completed in port or under way. Most exercises with equivalencies were completed under way, regardless of mission area.

In sum, the majority of exercises, including exercises with equivalencies that could be completed in port, were completed under way. The reasons for completing most exercises while under way may vary, but they include, at least, culture, tradition, or availability of underway time. It appears, however, that there are significant opportunities to complete more exercises in port.

Table 5.3 reprises the information presented previously regarding the location in which exercises with equivalencies were completed and presents, as well, analysis of where exercises that COMNAVSUR-FLANT N-7 believed could be completed in port were actually completed. As the data presented there indicate, 71 percent of the exercises that could be completed in port or under way were completed under way. Only 29 percent were completed in port. This observation suggests that current training practices do not take advantage of the possibility of completing exercises in port.

**Table 5.2**

**Where All Exercises Were Completed, by Mission Area**

| Mission Area | Exercises with Equivalencies | Total Completed | Total Completed in Port | Percentage Completed in Port |
|---|---|---|---|---|
| USW | 24 | 543 | 123 | 23 |
| AAW | 19 | 396 | 77 | 19 |
| SUW | 6 | 206 | 34 | 17 |

**Table 5.3**

**Where All Exercises Were Completed**

| | Exercises Completed | |
|---|---|---|
| Type of Exercise | Under Way (%) | In Port (%) |
| All exercises[a] | 77 | 23 |
| Exercises with equivalencies | 80 | 20 |
| Exercises that could be completed under way or in port | 71 | 29 |
| Exercises that could only be completed under way | 92 | 8 |

[a]All figures for nondeployed ships only. For deployed ships, 99 percent of all exercises were completed under way.

Our analysis indicates that there was considerable variation across mission areas in the extent to which exercises that could be completed in port were completed in port. There are eight mission areas in which a majority of exercises could be completed in port. The proportion of such exercises ranged from 7 percent for C2W exercises to 61 percent for noncombat operations (NCO). (See Appendix C for more details regarding the proportion of potential in-port exercises that were completed in port, organized by mission area.)

As was the case for exercises with equivalencies, it appears that more "could be completed under way or in port" exercises could, in fact, be completed in port. Because readiness standards are the same regardless of whether training is completed in port or under way, it seems unlikely that increasing the proportion of training exercises conducted in port would affect readiness.

Table 5.3 also indicates that 8 percent of "underway only" exercises were completed in port. This finding might be a result of reporting error or might suggest that exercises that were categorized as "must be completed under way" are, in fact, sometimes completed in port.

Table 5.4 summarizes the results of our analyses of where exercises that could be completed in port were actually completed.

The first column indicates the proportion of exercises that could be completed in port that have equivalencies and the proportion that do not. The data in this column reflect 6,756 exercises that were completed during the period we examined. Of these potential in-port exercises, 21 percent have equivalencies, and 79 percent do not.

In the second column, we present the percentage of each kind of potential in-port exercise that was actually completed in port. Exercises with equivalencies actually completed in port constitute 4 percent of those that could have been completed in port; exercises with no equivalencies that could be completed in port or under way and were actually completed in port constitute 22 percent of potential in-port exercises. **In sum, only 26 percent of all required exercises were completed in port—about one-fourth of the exercises that could have been completed in port.**

The set of exercises with no equivalencies that could have been completed under way or in port is the largest set of exercises in our data set (5,125 exercises). That only 22 percent of this large set of

**Table 5.4**

**Proportion of All Exercises That Could Be Completed in Port Actually Completed in Port**

| Exercise Type | Exercises That Could Be Completed in Port | |
|---|---|---|
| | Possible (%) | Actual (%) |
| Exercises with equivalencies | 21 | 4 |
| Exercises with no equivalencies that could be completed in port or under way | 79 | 22 |
| Total | 100 | 26 |

exercises was completed in port suggests that there is substantial opportunity for increasing the proportion of training exercises conducted in port and, by extension, for increasing the use of simulation in training.

# Chapter Six

# RECOMMENDATIONS

In this chapter, we present recommendations based on the quantitative analyses we conducted, the results of our interviews, and our review of other research on this topic. In presenting these recommendations, our aim is to provide information that can help to

- guide decisions about whether to increase the use of virtual and live simulation in surface force training

- identify mission areas and kinds of exercises that would be appropriate targets for increasing the use of simulation, especially virtual simulation

- develop strategies for purchasing and implementing simulators.

## DEFINE THE GOALS OF TRAINING

One of the great strengths of the Navy is the sharing of information between crew members and their ability to train themselves through this process of information sharing. The Navy trains under way and conducts its mission under way and forward deployed, which provides rich opportunities to support the training environment. This training method has been described by the Navy as "training the way we fight and fighting the way we train." The success of this method remains unchallenged.

Nonetheless, it is possible, even likely, that, for some purposes, in-port training on simulators would be as effective as underway training would be or, perhaps, even more effective in terms of producing proficiency (through more repetitions) and could also be more cost-effective. To determine the future balance of training, however—i.e., the roles of live and simulated training in relation to attaining performance goals—it is important to establish the goal of training. If the goal is to achieve the greatest proficiency, more resources have to be expended or significant process changes need to be made. If the goal is to reduce cost while maintaining the same proficiency, the trade-offs may be different.

Determining goals for training and whether simulation should be an important part of achieving them will require overcoming the cultural bias toward conducting training events under way. Shifting from a "the way we've always done it" stance to a "finding the best way" stance would enable the Navy to base its training goals on outcomes rather than with the means of achieving them.

## SPECIFY MEASURES OF EFFECTIVENESS FOR TRAINING

The Navy must decide how it wants to measure readiness through the completion of training events. A combination of readiness standards is currently being used in surface force training readiness. FXP exercise reporting is used for SORTS training readiness reporting, while ships are being trained and assessed to Navy Mission Essential Task Lists training requirements at the intermediate and advanced levels of training. Defining performance standards in a dynamic operational setting is challenging, but a clear set of standards that covers all phases of training across mission areas may lead to a clear sequencing of training, as well as to an efficient way of achieving training readiness.

The Navy must also determine the relationship between the use of simulators for training and proficiency. Some simulators do not provide the same level of fidelity as live events. This fidelity may affect the proficiency of the operator. To ascertain whether training on simulators produces a satisfactory level of proficiency, a metric for proficiency that is acceptable to Navy leaders must be established. Given such measures, the Navy can determine whether a shift toward greater use of simulation is warranted.

Measuring the relative effectiveness of in-port and underway training is difficult at best, but if the Navy is to pursue an increased use of simulation to replace live underway training, the combination of live (under way) and simulation (in port or under way) training to achieve maximum readiness must be determined. The configuration of live and simulated exercises is likely to be different for each mission area and, perhaps, for specific exercises within mission areas.

Assessments of levels of proficiency attained through simulators should include exercises for which underway training is believed to be necessary. Determining whether the initial stages of exercises that can only be done under way (e.g., engineering drills) can be simulated effectively may help improve the efficiency of these exercises when completed under way.

## USE MULTIPLE APPROACHES TO REDUCE UNDERWAY TRAINING

The majority of required FXP exercises need not be completed under way. The locations in which these exercises are completed do not affect a ship's training readiness rating in SORTS. That is, where a ship takes credit for an exercise, whether in port or under way, has no bearing on the ship's readiness rating. Thus, to increase the efficiency of underway training, the Navy should focus on using underway time for exercises that can only be done under way.

In addition to narrowing the set of exercises done under way, attention should be given to the way that exercises are organized and sequenced. In particular, "underway only" exercises should be prioritized, and high-priority exercises should be completed first. In addition, to achieve maximum training benefit from time at sea, exercises should be sorted into groups that can be completed simultaneously. Core engineering drills, for instance, would fall into this category. Other exercises could be completed on a "not to interfere" basis.

Training equivalencies should be used to their maximum extent. Whether because of custom, convention, or policy, exercises with equivalencies are frequently completed under way. It is up to the type commanders and numbered fleet commanders to assess the relative value of in-port and underway training, but the data presented here indicate that more exercises with equivalencies could be completed in port.

In addition, it would be useful to identify exercises for which equivalencies could be approved, thus expanding the range of exercises for which credit toward mission readiness could be earned by completing exercises in port. To the extent that exercises can be completed in port, underway time could be decreased or used more effectively.

## DEVELOP A SIMULATION STRATEGY

To expand the use of simulation in a way that optimizes both proficiency and the use of training resources, it is important to develop a simulation strategy—to determine whether, how, and for what simulation can be used. Here, we present several elements that might constitute such a strategy.

First, the Navy needs to clarify responsibility and authority for decisionmaking with regard to the use of simulation in training.

Second, simulation must be permissible under the SURFTRAMAN. Currently, training exercises must be completed on the ship's own equipment to earn credit in assessments of readiness. This policy is inconsistent with the idea of increasing the use of simulation for training. Moreover, it is inconsistent with the Navy's current practice of rotating crews to ships other than those on which they trained. The other military organizations we studied allow simulation on centrally located equipment, and this training is counted in assessments of the unit's operational readiness.

Third, the Navy should select areas for simulation in which simulation will provide the greatest benefit. In the next subsection, we identify kinds of exercises that could, potentially, be simulated.

### Exercises for Which Actions Do Not Depend on the Location of the Ship

The training that can be completed in port depends on the training requirement. For instance, it may be more efficient for individuals and teams to conduct in-port exercises for which the actions taken or the reactions to stimuli are known not to depend on the location of the ship. One such area is ASW. An ASW team can accomplish approximately the same level of training when the ship is tied to the pier, with external signals stimulating equipment. These benefits can extend to other mission areas as well.

## High-Frequency Exercises

The relationship between training methods and proficiency depends on the characteristics and complexity of the training exercise. Logically, it seems that exercises that require repetition would be good candidates for simulation because the consistency of the training environment would allow users to develop and refine their skills without the intrusion of irrelevant factors that may undermine performance.

Completing high-frequency exercises through simulation also has the potential to reduce costs. In the past, decisions about the selection of exercises for simulation appear to have been made without regard to where the greatest cost reductions could be achieved. Currently, one-third of the longest duration exercises (24 months) can be simulated, but only about one-tenth of the shortest duration exercises (three months or less) can be.

Making the simple assumption that costs per exercise are equal, focusing efforts to develop new simulators on exercises that require frequent repetition would provide a greater return on investment. Of course, the cost of exercises varies substantially, so determining which exercises to simulate would require a more-detailed financial analysis. Nevertheless, it seems logical that simulating high-frequency exercises would be more likely to be cost-effective than simulating low-frequency exercises and that the financial benefits of simulating high-frequency exercises could be maximized by focusing development efforts on the most costly high-frequency exercises.

## Exercises in Nontactical Mission Areas

Currently, all approved equivalencies are in tactical mission areas, but training for many nontactical missions could take place in port. Our analysis of the use of simulation by other militaries and in the private sector indicates that engineering, ship-handling functions, and maintenance exercises are all likely candidates for simulation. To the extent that such exercises could be simulated, it may be possible to reduce training costs or, at least, to increase the cost-effectiveness of training.

## Exercises Simulated by Other Military Organizations and the Private Sector

As we have shown, other military organizations, commercial aviation, and commercial shipping use simulation for training to a much greater extent than does the U.S. Navy, with good results. Indeed, in commercial shipping, almost all training takes place through simulation, and, in the Canadian Navy, underway training is reserved for exercises that involve multiple ships or live firing. Under Task Force Excel, the U.S. Navy is moving to align its qualification standards with those of the merchant marine. As this process goes forward, it should focus not only on standards but also on methods of meeting those standards. In defining those methods, the use of simulation for training should play a key role.

## DEVELOP AN INVESTMENT STRATEGY

The Navy should invest in the simulators that afford the best fidelity and should maximize their availability. Heretofore, limited availability of simulators with good fidelity has hindered the expansion of training through simulation.

Because multiple simulators are available for the same exercises, it appears that the direction of simulation has been to improve the fidelity of what has already been simulated. Although fidelity is, of course, important, it may be that, in some areas, fidelity is already satisfactory and that developing simulators in new mission areas should receive more attention.

## EXAMPLE: USING SIMULATION IN CASUALTY-CONTROL DRILLS

To demonstrate the feasibility of implementing our recommendations, we will present a brief example of how simulation might be used in engineering drills.

To date, the Navy's simulation capability, through the use of preapproved exercise equivalencies, has focused on the primary tactical warfare mission areas of AAW; administrative support unit; USW; C2W; and, to some extent, navigation. In other mission areas, outside of the schoolhouses, little simulation is used in fleet concentration areas, such as Norfolk, Virginia, and San Diego, California.

Engineering exercises, however, are manpower intensive and time-consuming and must be completed for every watch section before being reported in SORTS. These exercises consist of core and other casualty-control drills. Our research indicated that, although a majority of engineering exercises could be completed in port, more than 80 percent of the engineering exercises that could be completed in port were, in fact, completed under way.

As ships enter the basic phase of training, many resources (ATG and other engineering trainers) are devoted to increasing the proficiency of the engineering team. Some engineering drills can only be completed under way, but many can be completed in port. While engineering simulators, such as the DDG-51 Propulsion Trainer, are used at schoolhouses (e.g., Surface Warfare Officer School) to train personnel in casualty-control procedures, simulators of engineering central control stations are not available at fleet concentration centers.

However, the benefits of off-ship casualty-control drills are numerous and include

- improving engineering readiness
- increasing the realism of drills
- reducing the number of personnel required to complete the drill
- providing the capability to train while maintenance is being performed
- allowing more crew members opportunities for training in casualty-control procedures, including junior personnel, potential engineering officers of the watch, and other watch standers.

Of course, there are also drawbacks to simulating these training exercises. A simulated exercise is not, by definition, performed on a ship's own equipment. Further, substantial costs would be involved in purchasing new simulators, and a new off-ship training requirement might overburden the already full pierside training schedule.

We believe, however, that these disadvantages are outweighed by the advantages. Making a DDG-51 propulsion trainer available at fleet concentration areas would enable teams to practice ashore, thus increasing the efficiency of underway training and, potentially, reducing underway training requirements. Such a facility could also help train engineering watch sections for rotation to forward-deployed ships; allow more effective use of underway time; and, through practice, maintain currency of casualty-control procedures.

Implementing this procedure would, of course, require purchasing the trainer and developing procedures for its use. However, at least equally important, it would require a policy change, in that the type commander would have to change the SURFTRAMAN requirement that, to earn credit, training exercises must be done on a ship's own equipment.

## SUMMARY

The primary purposes of this investigation were to describe current training practices in the U.S. Navy, determine whether they revealed opportunities to increase the use of simulation in training, and develop recommendations that might inform decisions regarding training policies and practice.

In the course of our analysis, we drew on (1) Navy training documents, training reports, and employment schedules; (2) discussions with Navy fleet training authorities; (3) training authorities in other military and commercial organizations; and (4) the results of previous RAND research on the use of simulation for training in aviation.

The first part of our analysis revealed considerable complexity in the training architecture for the Navy's surface forces, as well as in the policies and funding patterns associated with the use of simulation. In our view, creating a unified policy on the use of simulation for training will require clarifying lines of authority and responsibility for the development, governance, and use of simulators. Such actions seem warranted by the Navy's interest in reducing training costs by augmenting or replacing live training with simulation.

The results of our empirical analysis, which focused on training for the DDG-51 ship class, indicated that there are numerous opportunities to increase in-port training through the use of simulation. After identifying exercises with equivalencies and distinguishing exercises that could only be completed under way from those that could be completed under way or in port, we used Navy training reports and employment schedules to determine which exercises were conducted under way and which were conducted in port. We found that the vast majority of exercises—including those with approved equivalencies and those that could be completed in port—were, in fact, completed under way. This outcome clearly indicates that, when judged by its own published standards and the criteria suggested by training experts, simulation is underused in training the surface force.

This conclusion is supported by our analysis of the use of simulation for training in other organizations. We examined the training practices of the British Navy, the Canadian Navy, commercial aviation, and commercial shipping. In all these organizations, simulation is used extensively for training. In military organizations, the exercises for which simulation is not used are those that involve live firing and multiple ships. In some cases, such as with the French MPA, simulation is used for practice and live exercises are used for evaluation. As the U.S. Navy changes its training requirements to be more closely aligned with those of the merchant marine, it may also be useful to adopt some of its training methods.

We do not, of course, mean to suggest that underway training can or should be eliminated. Some exercises will always require underway training, and, for others, proficiency might best be achieved though a combination of simulation and underway training. Thus, as the Navy moves to refine its training practices, it should carefully consider selecting exercises that can be simulated without reducing proficiency or readiness. Our analysis indicates that such exercises include (1) exercises for which there are approved equivalencies, (2) exercises that require frequent repetition, (3) exercises that do not involve live firing, and (4) exercises that are currently simulated by other military and commercial organizations.

Moving in these directions will require several changes. First, the requirement for completing exercises on a ship's own equipment to earn credit in the assessment of readiness must be modified. In addition, better-developed readiness standards will need to be adopted that provide criteria for use in developing simulators and evaluating their effectiveness. Finally, and perhaps most challenging, Navy culture will have to change so that training completed through simulation is accorded the same respect as training completed under way.

# COMNAVSURFLANT/COMNAVSURFPAC–
# SPECIFIED FXP EXERCISES

The table below lists all FXP exercises for the DDG-51 ship class, grouped by mission area. The numbers in the second column indicate the intervals at which M-ratings degrade if the exercise is not completed. For example, if the numbers are 3, 6, and 9, the M-rating degrades to M-2 if the exercise is not completed in three months, to M3 if the exercise is not completed in six months, and to M-4 if the exercise is not completed in nine months. Exercises with no value in the periodicity column need only be completed once during the IDTC. In general, the periodicities indicated here are based on the SURFTRAMAN issued in 1999. Guidance for a small subset of the exercises, however, is contained in SURFTRAMAN Bulletins. For more information on the governance of exercises, see the table notes at the end of this appendix.

In the second column, a $Y$ indicates that the exercise has an authorized equivalency, which means that a ship can earn credit toward its readiness rating by completing the exercise either under way or in port.

The data in the last column refer to exercises with no equivalencies. A $Y$ indicates that, according to the judgments of SURFLANT training officials, the exercise must be completed under way. Blank cells indicate that the exercise could be done under way or in port.

**Table A.1**

**FXP Exercises by Periodicity, Equivalency, and Location for Completion as Specified by COMNAVSURFOR**

| Mission Area | Periodicity (in months) | Authorized Equivalency | No Equivalency; Must Be Completed Under Way[a] |
|---|---|---|---|
| **AAW** | | | |
| AAW-10-I<br>COORD CAP/MSL EMPL | | Y | |
| AAW-11-SF<br>SUBSONIC ASMD RAID (FIRING) | | | Y |
| AAW-11-I<br>COORD CAP/MSL EMPL IN ECM ENVIRON | | Y | |
| AAW-12-SF<br>AA GUNNERY | | | Y |
| AAW-13-I<br>CINTEX | | Y | |
| AAW-14-I A/C<br>CONTROL/ASM INTERCEPT | | Y | |
| AAW-15-SF<br>INFO PROCEDURES | | Y | |
| AAW-16-SF<br>LIVE AAWEX | | | |
| AAW-17-SF<br>LINK 11 INTRUSION—JAMMING | | Y | |
| AAW-20-SF<br>CIWS READINESS EVAL | | | Y |
| AAW-21-SF<br>CIWS FIRING | | | Y |
| AAW-24-SF<br>DTE SEQUENCE (NONFIRING) | | | |
| AAW-26-SF<br>LINK 4A AIC | | | |
| AAW-27-SF<br>SUPERSONIC ASMD (FIRING)—LOW ALT | | | Y |
| AAW-2-SF<br>LINK 11 OPS | | Y | |

## Table A.1—Continued

| Mission Area | Periodicity (in months) | Authorized Equivalency | No Equivalency; Must Be Completed Under Way[a] |
|---|---|---|---|
| AAW-3-I | | | |
| AIC | | Y | |
| AAW-3-SF | | | |
| RADAR/IFF TRACKING | 3, 6, 9 | Y | |
| AAW-4-I | | | |
| LOST PLANE HOMING | | Y | |
| AAW-4-SF | | | |
| AA TGT DESIG AND ACQUISITION (NONFIRING) | | Y | |
| AAW-5-I AA | | | |
| TGT DESIG/ACQ IN MUL TGT ENV-CAP COORD | | Y | |
| AAW-6-SF | | | |
| S/S AIR TARGET DETECTION, TRACK, DESIG &ACQ | | Y | |
| AAW-7-I | | | |
| ECCM-CAP COORD IN MECH JAMMING | | Y | |
| AAW-7-SF | | | |
| TACTICAL AAW | 6, 12, 18 | Y | |
| AAW-8-I | | | |
| TAC AAW CAP MSL COORD | | Y | |
| AAW-9-I | | | |
| TAC AAW CAP/MSL COORD WITH COUNTERMEASURES | | Y | |
| **Total  AAW** | | 17 | 5 |
| **AMW** | | | |
| AMW-2/3-SF | | | |
| NSFS QUAL | | | Y |
| **Total AMW** | | | 1 |
| **USW** | | | |
| ASW-11-SF | | | |
| UNIDENT CONTACT REPORTING | 3, 6, 9 | Y | |
| ASW-13-SF | | | |
| PASSIVE TRACKING SHORT RANGE | | Y | |

## Table A.1—Continued

| Mission Area | Periodicity (in months) | Authorized Equivalency | No Equivalency; Must Be Completed Under Way[a] |
|---|---|---|---|
| ASW-14-SF ASW SEARCH | | Y | |
| ASW-15-SF SUB FAM | | | Y |
| ASW-18-SF SVTT (FIRING) | 6, 12, 18 | | Y |
| ASW-19-SF RTT (FIRING) | 6, 12, 18 | | Y |
| ASW-1-SF SVTT LOADING | | | |
| ASW-23-SF ASW A/C VECTAC (SIM) | 6, 12, 18 | Y | |
| ASW-24-SF LAMPS DROP CAP | 6, 12, 18 | | Y |
| ASW-28-SF CZ-BB OPS | 6, 9, 12 | Y | |
| ASW-29-SF INTERMEDIATE CONTACT MGMT | 6, 9, 12 | Y | |
| ASW-2-SF SONAR CASUALTY DRILL | 3, 6, 9 | | |
| ASW-31-SF CLS IN SRCN SURV FORCE | | Y | |
| ASW-32-SF PERIMETER SCRN SURF FORCE | | Y | |
| ASW-33-SF BARRIER SEARCH/DEFEND OBJ AREA | | Y | |
| ASW-35-SF COOR ATK W/EVASION | | Y | |
| ASW-38-SF CZ-EX PASS BUOY | 6, 12, 18 | Y | |
| ASW-3-SF BASIC CONTACT MGMT | | Y | |
| ASW-40-SF HELO CONT ASW SCREEN | | | |
| ASW-41-SF LAMPS III HELO CONTROL | | Y | |

## Table A.1—Continued

| Mission Area | Periodicity (in months) | Authorized Equivalency | No Equivalency; Must Be Completed Under Way[a] |
|---|---|---|---|
| ASW-42-SF SHIP/FIX WNG A/C CONTROL | 6, 9, 12 | Y | |
| ASW-43-SF LAMPS III/SHIP ATTACK | | Y | |
| ASW-5-I SHALLOW WATER | 3, 6, 9 | | |
| ASW-5-SF OWNSHIP ACOUSTIC SIGN | 6, 12, 18 | | Y |
| ASW-6-SF ACOUSTIC ENV PRED | 3, 6, 9 | | |
| ASW-8-I CHOKE POINT XST[b] | 24 | | |
| ASW-8-SF ACTIVE TRACKING | | Y | |
| ASW-9-SF ACT MULTI-MODE LNG RN | | Y | |
| CONTACT_ACOUSTIC TIME-ANALYSIS[c] | | | |
| CONTACT_ACTIVE TIME-ACTIVE SENSORS[c] | | | |
| CONTACT_LT TIME-LIVE TARGET[c] | | | Y |
| CONTACT_PA TIME-PASSIVE SENSORS[c] | | | |
| **Total USW** | | 17 | 6 |
| **C2W** | | | |
| C2W-10-SF COORD MULTI SHIP EW | | | Y |
| C2W-11-SF CHAFF FIRING | | | Y |
| C2W-12-SF LAMPS MK III U/W DEM | | Y | |
| C2W-13-SF MISSILE/THREAT ELECTONIC ATTACK | | | Y |

**Table A.1—Continued**

| Mission Area | Periodicity (in months) | Authorized Equivalency | No Equivalency; Must Be Completed Under Way[a] |
|---|---|---|---|
| C2W-14-SF EW ASSESSMENT | | | |
| C2W-15-SF MK36 DECOY LOADEX | 6, 12, 18 | | |
| C2W-16-SF COORD CHAFF FIRING[d] | 12, 18, 24 | | Y |
| C2W-2-SF ES DETECTION, ANALYSIS AND REPORTING | 3, 6, 9 | Y | |
| C2W-30-SF DC&T EX | | | |
| C2W-31-SF TACTICAL COLLECTION, ANALYSIS AND REPORTING | 3, 6, 9 | | |
| C2W-33-SF TACTICAL AIR TARGET | | | |
| C2W-37-SF RDF SKYWAVE/GROUND PRESENTATION EXER | | | Y |
| C2W-3-SF EXTENDED EMCON | 3, 6, 9 | | Y |
| C2W-4-SF EMCON SET & MODIFICATION | 3, 6, 9 | | |
| C2W-5-SF SAT VULNERABILITY | 3, 6, 9 | | |
| C2W-6-SF ES WATCH EVAL | 3, 6, 9 | | |
| C2W-7-SF COMP EW EX PH I | | | |
| C2W-8-SF COMP EW EX PH II | | | |
| C2W-9-SF COMP EW EX PH III | | | Y |
| Total C2W | | 2 | 7 |

## Table A.1—Continued

| Mission Area | Periodicity (in months) | Authorized Equivalency | No Equivalency; Must Be Completed Under Way[a] |
|---|---|---|---|
| **CCC** | | | |
| CCC DATE LAST U/W OVERNIGHT | Report as occurs | | Y |
| CCC-10-SF FLASHING LIGHT PROCEDURES | 3, 6, 9 | | |
| CCC-11-SF SEMAPHORE | 3, 6, 9 | | |
| CCC-12-SF IMITATIVE DECEPTION AND JAMMING | 3, 6, 9 | Y | |
| CCC-13-SF EAP EMERG DESTRUCT CARDS | 3, 6, 9 | | |
| CCC-14-SF SYSCON-QMS | 3, 6, 9 | | |
| CCC-15-SF NTDS INITIATION AND OPS | 3, 6, 9 | | |
| CCC-16-SF AEGIS DOCTRINE MGMT | 3, 6, 9 | | |
| CCC-17-SF LINK 11 FAST FREQ CHANGE | 3, 6, 9 | | |
| CCC-18-SF TACINTEL COMM OPS | 3, 6, 9 | | |
| CCC-19-SF COMP COMMS ASSESSMENT | 12, 24, 36 | | |
| CCC-1-SF SYSCON-FLT BCST TYPE N | 3, 6, 9 | | |
| CCC-20-SF SYSCON-SI TERM/Z TERM | 9, 12, 18 | | |
| CCC-21-SF SYSCON-OPINTEL BCST/SI COMM (N SYS) | 6, 12, 18 | Y | |
| CCC-22-SF SYSCON-SPRAC NET | 6, 12, 18 | | |
| CCC-23-SF CRITIC HANDLING | 3, 6, 9 | | |
| CCC-24-SF SYSCOM_NB/WB SATCOM | 3, 6, 9 | | |

## Table A.1—Continued

| Mission Area | Periodicity (in months) | Authorized Equivalency | No Equivalency; Must Be Completed Under Way[a] |
|---|---|---|---|
| CCC-26-SF<br>SYSCON-EHF SATCOM | 3, 6, 9 | | |
| CCC-29-SF<br>OTCIXS/TADIXS SYSTEM | 3, 6, 9 | | |
| CCC-2-SF<br>COMM OP S PLANNING | 3, 6, 9 | | |
| CCC-30-SF<br>SYSCON-OTAT/OTAR | 3, 6, 9 | | |
| CCC-31-SF<br>SYSCON-NAVMACS II | 3, 6, 9 | | |
| CCC-3-SF<br>HELO ELVA CONTROL | 6, 12, 18 | | |
| CCC-4-SF<br>SYSCON-SHIP TERM EX FPR B. C, D & G SYS | 3, 6, 9 | | |
| CCC-5-SF<br>SYSCON-SECURE VOICE SYS | 3, 6, 9 | Y | |
| CCC-6-SF<br>R/T DRILLS | 3, 6, 9 | Y | |
| CCC-7-SF<br>TACTICAL MANEUVERS | 3, 6, 9 | | |
| CCC-8-SF<br>TTY CKT PROCEDURES | 3, 6, 9 | Y | |
| CCC-9-SF<br>FLAGHOIST SIGNAL PROCEDURES | 3, 6, 9 | | |
| **Total CCC** | | 5 | 1 |

**FSO**

FSO-M-10-SF
  SMOKE INHALATION

FSO-M-11-SF
  BURNS

FSO-M-1-SF
  BTL DRESSING STATIONS

FSO-M-2-SF
  CASUALTY TRANSPORT

## Table A.1—Continued

| Mission Area | Periodicity (in months) | Authorized Equivalency | No Equivalency; Must Be Completed Under Way[a] |
|---|---|---|---|
| FSO-M-3-SF<br>  COMPOUND FRACTURES | | | |
| FSO-M-4-SF<br>  SUCKING CHEST WOUND | | | |
| FSO-M-5-SF<br>  ABDOMINAL WOUNDS | | | |
| FSO-M-6-SF<br>  AMPUTATION | | | |
| FSO-M-7-SF<br>  FACIAL WOUND | | | |
| FSO-M-8-SF<br>  ELECTRIC SHOCK | | | |
| FSO-M-9-SF<br>  MASS CASUALTY | | | |
| **Total FSO** | | 0 | 0 |
| **INT** | | | |
| INT-1-SF(MS)<br>  INTEL COLL & RPTG TEAM | | | |
| INT-1-SF(OP)<br>  OPINTEL DATA COLL | | | |
| INT-1-SF(RP)<br>  INTEL RPTG LOCATORS | | | |
| INT-2-SF(OP)<br>  OPINTEL PLOT & BRIEF | | | |
| INT-2-SF(RP)<br>  INTEL REPTG - IIR | | | |
| INT-3-SF(OP)<br>  C2W/INFO WARFARE CONN | | | |
| INT-4-SF(RP)<br>  SURVINTCOLEX | | | |
| INT-5-SF(RP)<br>  INCSEA/DANGER MIL ACTS<br>  EXERCISE | | | |
| **Total INT** | | 0 | 0 |

**Table A.1—Continued**

| Mission Area | Periodicity (in months) | Authorized Equivalency | No Equivalency; Must Be Completed Under Way[a] |
|---|---|---|---|
| **MOB–Engineering** | | | |
| MBGGM– | | | |
| CLASS B FIRE GTG | 3, 6, 12 | | |
| MBGTM– | | | |
| CLASS B FIRE GTM MODUL | 3, 6, 12 | | |
| MCASF–GT | | | |
| COOL AIR FAILURE | 3, 6, 12 | | |
| MCBF– | | | |
| B FIRE IN MAIN SPACE | 3, 6, 12 | | |
| MCCFG– | | | |
| CLASS C FIRE GEN | 3, 6, 12 | . | |
| MCCFS– | | | |
| CLASS C FIRE SWB | 3, 6, 12 | | |
| MCFED– | | | |
| CLASS C FIRE EDS | 3, 6, 12 | | |
| MEPTV–PT | | | |
| VIBS HI GTM | 3, 6, 12 | | |
| MGGOS– | | | |
| GAS GEN OVERSPEED | 3, 6, 12 | | |
| MGGS– | | | |
| GAS GEN STALL GTM | 3, 6, 12 | | |
| MGHIT– | | | |
| HIGH GT INLET TEMP | 3, 6, 12 | | |
| MHBGTG– | | | |
| HOT BEARING GTG | 3, 6, 12 | | |
| MHBRG– | | | |
| HOT BEARING MRG | 3, 6, 12 | | Y |
| MHLSB– | | | |
| HOT LINE SHAFT BEARING | 3, 6, 12 | | Y |
| MHTIT–PT | | | |
| INLET TEMP HI GTM | 3, 6, 12 | | |
| MLCRP– | | | |
| LOSS CPP PITCH CONTROL | 3, 6, 12 | | Y |
| MLCWS– | | | |
| LOSS CHILL WATER | 3, 6, 12 | | |
| MLEPC– | | | |
| LOSS OF EPCC | 3, 6, 12 | | |

## Table A.1—Continued

| Mission Area | Periodicity (in months) | Authorized Equivalency | No Equivalency; Must Be Completed Under Way[a] |
|---|---|---|---|
| MLFOP– LOSS MN ENG F/O PRESS | 3, 6, 12 | | |
| MLGGO– GTG LOW L/O PRESS | 3, 6, 12 | | |
| MLHOL– LEAK CRP/CPP SYSTEM | 3, 6, 12 | | Y |
| MLHOP– LOSS CRP/CPP PRESSURE | 3, 6, 12 | | Y |
| MLLOL– MJ L/O LEAK MN ENG/MRG | 3, 6, 12 | | Y |
| MLLOPR– LO LOP REDUCTION GEAR | 3, 6, 12 | | Y |
| MLPACC– LOSS OF PACC CONSOLE | 3, 6, 12 | | |
| MLPLA– LOSS PWR LEVEL ACCTR | 3, 6, 12 | | Y |
| MLPTO– LOW LOP GTM | 3, 6, 12 | | |
| MLSC– LOSS STEERING CONTROL | 3, 6, 12 | | |
| MLSCU– LOSS SHAFT CONTRL UNIT | 3, 6, 12 | | |
| MMF– FLOODING IN MAIN SPACE | 3, 6, 12 | | |
| MMFOL– MAJOR FUEL OIL LEAK | 3, 6, 12 | | |
| MNVGG– GTG NOISE/VIBRATION | 3, 6, 12 | | |
| MOSGG– OVERSPEED SSGTG | 3, 6, 12 | | |
| MPSFG– GTG POST SHUTDOWN FIRE | 3, 6, 12 | | |
| MPSFP– POST SHUTDOWN FIRE | 3, 6, 12 | | |
| MPTOS– PWR TURB OVERSPEED | 3, 6, 12 | | |

**Table A.1—Continued**

| Mission Area | Periodicity (in months) | Authorized Equivalency | No Equivalency; Must Be Completed Under Way[a] |
|---|---|---|---|
| MNVRG– NOISE/VIB MRG/SHAFT | 3, 6, 12 | | |
| **Total MOB–Engineering** | | 0 | 8 |
| **MOB–Damage Control** | | | |
| MOB-D-10-SF RESCUE/ASSISTANCE IN PORT | 3, 6, 9 | | |
| MOB-D-11-SF SETTING MATL COND: PHASE 1:YOKE, PHASE 2, ZEBRA | 3, 6, 12 | | |
| MOB-D-12-SF U/W HULL DAMAGE PH 1 AND 2 | 3, 6, 12 | | |
| MOB-D-13-SF SHORING | 3, 6, 9 | | |
| MOB-D-14-SF FIRE EXTINGUISHING/ SMOKE CLEARING | 1, 2, 3 | | |
| MOB-D-15-SF CHEMICAL ATTACK | 6, 12, 18 | | |
| MOB-D-20-SF ISOLATE/PATCH DAMAGED PIP | 3, 6, 12 | | |
| MOB-D-21-SF MAJOR FLOOD MAIN PROPULSION SPACE | 3, 6, 12 | | |
| MOB-D-23-SF LOCATING DC FITTINGS | 6, 12, 18 | | |
| MOB-D-24-SF DARKEN SHIP | 6, 12, 18 | | |
| MOB-D-27-SF HELO CRASH F/F | , 2, 3 | | |
| MOB-D-2-SF RELIEF OF VITAL STATIONS | 3, 6, 12 | | |
| MOB-D-31-SF TOXIC GAS DRILL | 3, 6, 9 | | |
| MOB-D-3-SF MANNING BATTLE STATIONS | 1, 2, 3 | | |
| MOB-D-4-SF EMERG INTERIOR COMMS | 3, 6, 12 | | |

**Table A.1—Continued**

| Mission Area | Periodicity (in months) | Authorized Equivalency | No Equivalency; Must Be Completed Under Way[a] |
|---|---|---|---|
| MOB-D-5-SF | | | |
| TOPSIDE DAMAGE | 3, 6, 12 | | |
| MOB-D-7-SF | | | |
| PROV CASUALTY POWER | 12, 24, 36 | | |
| MOB-D-8-SF | | | |
| MAJOR CONFLAGRATION | 6, 9, 12 | | |
| MOB-D-9-SF | | | |
| MAIN PROPULSIONSPACE FIRE | 3, 6, 9 | | |
| **Total MOB–Damage Control** | | 0 | 0 |
| **MOB–Navigation** | | | |
| MOB-U/W | Report as | | |
| OVERNIGHT | occurs | | Y |
| MOB-N-1-SF | | | |
| NAV IN EW ENVIRONMENT | 6, 12, 18 | | Y |
| MOB-N-2-SF | | | |
| OPEN OCEAN NAV | 3, 6, 9 | | Y |
| MOB-N-4-SF | | | |
| PILOTING BY GYRO | 3, 6, 9 | | Y |
| MOB-N-5-SF | | | |
| PRECISION ANCHORING | 6, 12, 18 | | Y |
| MOB-N-6-SF | | | |
| LOW VIS PILOTING | 3, 6, 9 | | Y |
| MOB-N-7-SF | | | |
| PILOTING—LOSS OF GYRO | 3, 6, 9 | | Y |
| MOB-N-9-SF | | | |
| LOSS OF STEERING | 3, 6, 9 | | Y |
| **Total MOB–Navigation** | | 0 | 8 |
| **MIW–Mine Warfare** | | | |
| MIW-8.7-SF | | | |
| TRANS SWEPT CHANNEL | 3, 6, 9 | | Y |
| **Total MIW–Mine Warfare** | | 0 | 1 |
| **MOB–Seamanship** | | | |
| MOB-S-10-SF | | | |
| U/W FUEL-DAY | 6, 12, 18 | | Y |

**Table A.1—Continued**

| Mission Area | Periodicity (in months) | Authorized Equivalency | No Equivalency; Must Be Completed Under Way[a] |
|---|---|---|---|
| MOB-S-10-SF U/W FUEL-NIGHT | 6, 12, 18 | | Y |
| MOB-S-11-SF EM BREAKAWAY (DAY) | | | Y |
| MOB-S-12-SF TOW/BE TOWED | | | Y |
| MOB-S-13-SF HELO LAND/LAUNCH | 3, 6, 9 | | Y |
| MOB-S-14-SF SAREX | | | |
| MOB-S-15-SF HIFR | 3, 6, 9 | | Y |
| MOB-S-16-SF U/W PROV (DAY) | | | Y |
| MOB-S-18-SF GET U/W W/DUTY SEC | | | Y |
| MOB-S-25-SF A/C ON DECK REFUEL | 3, 6, 9 | | Y |
| MOB-S-2-SF HEAVY WEAX | | | |
| MOB-S-33-SF HOIST/LOWER BOATS | 3, 6, 9 | | |
| MOB-S-34-SF RESCUE SWIMMER | 3, 6, 9 | | |
| MOB-S-3-SF ANCHORING DAY | | | Y |
| MOB-S-4-SF MOORING TO A BUOY | 12, 18, 24 | | Y |
| MOB-S-5-SF MOOR TO PIER/SHIP AT ANCHOR | | | Y |
| MOB-S-6-SF MAN OVERBOARD DAY | 3, 6, 9 | | Y |
| MOB-S-7-SF PREPS-ABANDON SHIP | 6, 12, 18 | | |
| MOB-S-8-SF VERTREP | | | Y |

## Table A.1—Continued

| Mission Area | Periodicity (in months) | Authorized Equivalency | No Equivalency; Must Be Completed Under Way[a] |
|---|---|---|---|
| MOB-S-9-SF U/W TRANSFER (SYNTHETIC HIGHLINE) | 6, 12, 18 | | Y |
| **Total MOB–Seamanship** | | 0 | 15 |
| **NCO** | | | |
| NCO-11-SF COMBAT CLASS C FIRE | 3, 6, 9 | | |
| NCO-12-SF EQPT CAS REPAIR | 3, 6, 9 | | |
| NCO-15-SF ALTERNATE PWR SOURCE | 3, 6, 9 | | |
| NCO-16-SF ECC/ESS | | | |
| NCO-18-SF SECURITY DRILLS | 3, 6, 9 | | |
| NCO-19-SF SMALL ARMS QUALS | | | |
| NCO-1-SF PREP FOR ELEC SPACES | 3, 6, 9 | | |
| NCO-28-SF ROE | 1, 2, 3 | | |
| NCO-29-SF DEFENSE VS U/W SWMRS | 12, 18, 24 | | |
| NCO-30-SF SHIP PENETRATION—BASIC | 12, 18, 24 | | |
| NCO-32-SF TERRORIST A/C ATTACK | 6, 12, 18 | | |
| NCO-33-SF SMALL BOAT ATTACK | 6, 12, 18 | | |
| NCO-34-SF BOMB THREAT | 6, 12, 18 | | |
| NCO-35-SF HOSTAGE SITUATION | 6, 12, 18 | | |
| NCO-36-SF FLOATING DEVICES | 6, 12, 18 | | |

## Table A.1—Continued

| Mission Area | Periodicity (in months) | Authorized Equivalency | No Equivalency; Must Be Completed Under Way[a] |
|---|---|---|---|
| NCO-38-SF VBSS | 6, 12, 18 | | |
| NCO-39-SF ATFP PLANNING P/S[d] | 6, 12, 18 | | |
| NCO-40-SF ATFP PLAN EXEC P/S[d] | 6, 12, 18 | | |
| NCO-41-SF ATFP PLANNING W/S[d] | 6, 12, 18 | | |
| NCO-42-SF ATFP PLAN EXEC W/S[d, e] | 18, 24 | | |
| **Total NCO** | | 0 | 0 |
| **STW** | | | |
| STW-1-SF MISSION DATA UPDATE | 1, 2, 3 | | |
| STW-21-A TLAM/C LAUNCH (SIM) | 6, 12, 18 | | |
| **Total STW** | | 0 | 0 |
| **SUW** | | | |
| SUW-10-SF OTH-T | 6, 12, 18 | Y | |
| SUW-12-SF VISUAL IDENT COUNTER | 6, 12, 18 | | Y |
| SUW-13-SF ATTACK/REATTACK EXER FOR SSM SHIPS | 6, 12, 18 | Y | |
| SUW-14-SF SAG LAMPS TACTICS | 6, 12, 18 | Y | |
| SUW-17-SF HI SPD SURF ENGAGEMENT | 6, 12, 18 | | Y |
| SUW-18-SF DATABASE MGMT | 6, 12, 18 | | |
| SUW-19-SF HI SPD QUICK FIRE EX | | | Y |
| SUW-1-I OTH SURVEILLANCE, SEARCH & DETECTION | 6, 12, 18 | Y | |

## Table A.1—Continued

| Mission Area | Periodicity (in months) | Authorized Equivalency | No Equivalency; Must Be Completed Under Way[a] |
|---|---|---|---|
| SUW-1-SF COMBINED AIR/SURFACE TRACKING | | Y | |
| SUW-2-I SAG TACTICS W/FIX WING A/C SUPPORT | 6, 12, 18 | Y | |
| SUW-2-SF LONG RANGE PASSISVE TRACKING AND TARGETING | | Y | |
| SUW-3-I SUW FREEPLAY EX | 6, 12, 18 | | |
| SUW-5-SF HSMST | | | Y |
| SUW-7-SF ALT/LCL CTRL LONG RANGE FIRE, HI SPD TARGET | | | Y |
| SUW-9-SF SURF TRKG NTDS/AEGIS SLAMEX | 3, 6, 9 | | |
| **Total SUW** | | 7 | 5 |

[a]Categorization of exercises as "must be completed in port" are based on the judgments of SURFLANT training officials as reported in interviews.

[b]This exercise must be repeated every 24 months. If it is not repeated at this interval, the M-rating degrades to M-4.

[c]ASW training requirements are governed by SURFTRAMAN Bulletin NR 410. Contact time M-ratings are based on the number of hours (live or synthetic) accumulated over the past six months.

[d]Training requirements derived from COMNAVSURFORINST 3502.1.

[e]This exercise degrades to M-2 if not completed in 18 months and to M-4 if not completed in 24 months.

# COMPLETION OF EXERCISES
# WITH EQUIVALENCIES

### Table B.1

### Where Exercises with Equivalencies Were Completed

| | Where Exercise Was Completed | | |
| Mission Area | In Port | Under Way | Total |
|---|---|---|---|
| **AAW** | | | |
| AAW-10-I | | | |
| COORD CAP/MSL EMPL | 6 | 12 | 18 |
| AAW-11-I | | | |
| COORD CAP/MSL EMPL IN ECM | 6 | 7 | 13 |
| AAW-13-I | | | |
| CINTEX | 3 | 2 | 5 |
| AAW-14-I | | | |
| A/C CONTROL/ASM INTERCEPT | 2 | 8 | 10 |
| AAW-15-SF | | | |
| INFO PROCEDURES | 6 | 23 | 29 |
| AAW-17-SF | | | |
| LINK 11 INTRUSION-JAMMING | 0 | 7 | 7 |
| AAW-2-SF | | | |
| LINK 11 OPS | 8 | 63 | 71 |
| AAW-3-I | | | |
| AIC | 3 | 27 | 30 |
| AAW-3-SF | | | |
| RADAR/IFF TRACKING | 8 | 43 | 51 |
| AAW-4-I | | | |
| LOST PLANE HOMING | 2 | 7 | 9 |

## Table B.1—Continued

| Mission Area | Where Exercise Was Completed | | Total |
|---|---|---|---|
| | In Port | Under Way | |
| AAW-4-SF<br>AA TGT DESIG AND ACQUISITION<br>(NONFIREING) | 7 | 32 | 39 |
| AAW-5-I<br>AA TGT DESIG/ACQ IN MULT TGT ENV-<br>CAP COORD | 4 | 11 | 15 |
| AAW-6-SF<br>S/S AIR TARGET DETECTION, TRACK,<br>DESIG AND ACQ | 6 | 13 | 19 |
| AAW-7-I<br>ECCM-CAP COORD IN MECH JAMMING | 0 | 5 | 5 |
| AAW-7-SF<br>TACTICAL AAW | 9 | 32 | 41 |
| AAW-8-I<br>TAC AAW CAP/MSL COORD | 4 | 14 | 18 |
| AAW-9-I<br>TAC AAW CAP/MSL COORD WITH<br>COUNTERMEASURES | 3 | 13 | 16 |
| **Total AAW** | 77 | 319 | 396 |
| **USW** | | | |
| ASW-11-SF<br>UNIDENT CONTACT REPORTING | 8 | 37 | 45 |
| ASW-13-SF<br>PASSIVE  TRACKING SHORT RANGE | 11 | 43 | 54 |
| ASW-14-SF<br>ASW SEARCH | 12 | 48 | 60 |
| ASW-23-SF<br>ASW A/C VECTAC (SIM) | 7 | 26 | 33 |
| ASW-28-SF<br>CZ-BB OPS | 6 | 17 | 23 |
| ASW-29-SF<br>INTERMEDIATE CONTACT MGMT | 10 | 28 | 38 |
| ASW-31-SF<br>CLS IN SRCN SURV FORCE | 7 | 12 | 19 |
| ASW-32-SF<br>PERIMETER SCRN SURF | 7 | 10 | 17 |

## Table B.1—Continued

| Mission Area | Where Exercise Was Completed | | |
| --- | --- | --- | --- |
| | In Port | Under Way | Total |
| ASW-33-SF<br>    BARRIER SEARCH/DEFEND OBJ AREA | 6 | 15 | 21 |
| ASW-35-SF<br>    COORD ATK W/EVASIOND | 4 | 9 | 13 |
| ASW-38-SF<br>    CZ-EX PASS BUOY | 1 | 4 | 5 |
| ASW-3-SF<br>    BASIC CONTACT MGMT | 3 | 26 | 29 |
| ASW-41-SF<br>    LAMPS III HELO CONTROL | 8 | 23 | 31 |
| ASW-42-SF<br>    SHIP/FIX WNG A/C COR | 2 | 20 | 22 |
| ASW-43-SF<br>    LAMPS III/SHIP ATTACK | 4 | 7 | 11 |
| ASW-8-SF<br>    ACTIVE TRACKING | 18 | 55 | 73 |
| ASW-9-SF<br>    ACT MULTI-MODE LNG RN | 9 | 31 | 40 |
| **Total USW** | 123 | 411 | 534 |
| **C2W** | | | |
| C2W-12-SF<br>    LAMPS MK III U/W DEM | 1 | 6 | 7 |
| C2W-2-SF<br>    ES DETECTION, ANALYSIS AND<br>    REPORTING | 7 | 63 | 70 |
| **Total C2W** | 8 | 69 | 77 |
| **CCC** | | | |
| CCC-12-SF<br>    IMITATIVE DECEPTION AND JAMMING | 2 | 15 | 17 |
| CCC-21-SF<br>    SYSCON-OPINTEL BCST/SI COMM<br>    (N SYS) | 3 | 5 | 8 |
| CCC-5-SF<br>    SYSCON-SECURE VOICE SYS | 20 | 55 | 75 |
| CCC-6-SF<br>    R/T DRILLS | 17 | 57 | 74 |

**Table B.1—Continued**

| | Where Exercise Was Completed | | |
| Mission Area | In Port | Under Way | Total |
|---|---|---|---|
| CCC-8-SF | | | |
| TTY CKT PROCEDURES | 17 | 47 | 64 |
| **Total CCC** | 59 | 179 | 238 |
| **SUW** | | | |
| SUW-10-SF | | | |
| OTH-T | 8 | 36 | 44 |
| SUW-13-SF | | | |
| ATTACK/REATTACK EXER FOR SSM SHIPS | 7 | 17 | 24 |
| SUW-14-SF | | | |
| SAG LAMPS TACTICS | 1 | 10 | 11 |
| SUW-1-I | | | |
| OTH URVEILLANCE, SEARCH & DETECTION | 7 | 32 | 39 |
| SUW-1-SF | | | |
| COMBINED AIR/SURFACE TRACKING | 4 | 43 | 47 |
| SUW-2-I | | | |
| SAG TACTICS W/FIX WING A/C SUPPORT | 4 | 8 | 12 |
| SUW-2-SF | | | |
| LONG RANGE PASSIVE TRACKING AND TARGETING | 3 | 26 | 29 |
| **Total SUW** | 34 | 172 | 206 |
| **Grand Total** | 301 | 1,150 | 1,451 |

# COMPLETION OF EXERCISES
# WITHOUT EQUIVALENCIES

Table C.1

**Where Exercises Without Equivalencies That Could Be
Completed Under Way or in Port Were Completed**

| Mission Area | Where Exercise Was Completed | | |
| --- | --- | --- | --- |
| | In Port | Under Way | Total |
| **AAW** | | | |
| AAW-16-SF | | | |
| LIVE AAWEX | 0 | 2 | 2 |
| AAW-24-SF | | | |
| DTE SEQUENCE (NON-FIRING) | 6 | 40 | 46 |
| AAW-26-SF | | | |
| LINK 4A AIC | 2 | 8 | 10 |
| **Total AAW** | 8 | 50 | 58 |
| **USW** | | | |
| ASW-1-SF | | | |
| SVTT LOADING | 9 | 28 | 37 |
| ASW-2-SF | | | |
| SONAR CASUALTY DRILL | 12 | 38 | 50 |
| ASW-40-SF | | | |
| HELO CONT ASW SCREEN | 2 | 14 | 16 |
| ASW-5-I | | | |
| SHALLOW WATER | 3 | 11 | 14 |
| ASW-6-SF | | | |
| ACOUSTIC ENV PRED | 7 | 61 | 68 |

**Table C.1—Continued**

| Mission Area | In Port | Under Way | Total |
|---|---|---|---|
| | **Where Exercise Was Completed** | | |
| ASW-8-I | | | |
| CHOKE POINT XST | 1 | 12 | 13 |
| CONTACT | | | |
| ACOUSTIC TIME-ANALYSIS | 21 | 45 | 66 |
| CONTACT | | | |
| ACTIVE TIME-ACTIVE SENSORS | 19 | 44 | 63 |
| CONTACT | | | |
| PA TIME-PASSIVE SENSORS | 22 | 45 | 67 |
| **Total USW** | 96 | 298 | 394 |
| **C2W** | | | |
| C2W-14-SF | | | |
| EW ASSESSMENT | 1 | 1 | 2 |
| C2W-15-SF | | | |
| MK 36 DECOY LOADEX | 4 | 34 | 38 |
| C2W-30-SF | | | |
| DC&T ANALYSIS AND REPORTING | 1 | 5 | 6 |
| C2W-31-SF | | | |
| TACTICAL COLLECTION, ANALYSIS AND REPORTING | 0 | 8 | 8 |
| C2W-33-SF | | | |
| TACTICAL AIR TARGET | 1 | 15 | 16 |
| C2W-4-SF | | | |
| EMCON SET & MODIFICATION | 4 | 61 | 65 |
| C2W-5-SF | | | |
| SATELLITE VULNERABILITY | 1 | 24 | 25 |
| C2W-6-SF | | | |
| ES WATCH EVAL | 4 | 45 | 49 |
| C2W-7-SF | | | |
| COMP EW EX PH I | 0 | 17 | 17 |
| C2W-8-SF | | | |
| COMP EW EX PH II | 1 | 8 | 9 |
| **Total C2W** | 17 | 218 | 235 |
| **CCC** | | | |
| CCC-10-SF | | | |
| FLASHING LIGHT PROCEDURES | 13 | 53 | 66 |
| CCC-11-SF | | | |
| SEMAPHORE | 9 | 50 | 59 |

**Table C.1—Continued**

| Mission Area | Where Exercise Was Completed | | |
| | In Port | Under Way | Total |
| --- | --- | --- | --- |
| CCC-13-SF | | | |
| EAP EMERG DESTRUCT CARDS | 15 | 17 | 32 |
| CCC-14-SF | | | |
| SYSCON-QMS | 11 | 25 | 36 |
| CCC-15-SF | | | |
| NTDS INITIATION & OP | 6 | 55 | 61 |
| CCC-16-SF | | | |
| AEGIS DOCTRINE MGMT | 9 | 39 | 48 |
| CCC-17-SF | | | |
| LINK 11 FAST FREQ CHANGE | 9 | 44 | 53 |
| CCC-18-SF | | | |
| TACINTEL COMM OPS | 1 | 13 | 14 |
| CCC-19-SF | | | |
| COMP COMM ASSESSMENT | 8 | 9 | 17 |
| CCC-1-SF | | | |
| SYSCON-FLT BCST TYPE N | 20 | 49 | 69 |
| CCC-20-SF | | | |
| SYSCON-SI TERM TTY Z | 1 | 3 | 4 |
| CCC-22-SF | | | |
| SYSCON-SPRAC NET | 1 | 3 | 4 |
| CCC-23-SF | | | |
| CRITIC HANDLING | 0 | 8 | 8 |
| CCC-24-SF | | | |
| SYSCOM NB/WB SATCOM | 17 | 49 | 66 |
| CCC-26-SF | | | |
| SYSCON-EHF SATCOM | 13 | 48 | 61 |
| CCC-29-SF | | | |
| OTCIXS/TADIXS SYSTEM | 9 | 50 | 59 |
| CCC-2-SF | | | |
| COMM OP S PLANNING | 20 | 44 | 64 |
| CCC-30-SF | | | |
| SYSCON OTAT/OTAR | 19 | 53 | 72 |
| CCC-31-SF | | | |
| SYSCON NAVMACS II | 15 | 41 | 56 |
| CCC-3-SF | | | |
| HELO ELVA CONTROL | 1 | 10 | 11 |

**Table C.1—Continued**

| Mission Area | Where Exercise Was Completed | | Total |
|---|---|---|---|
| | In Port | Under Way | |
| CCC-4-SF<br>SYSCON-SHIP TERM EX FPR B, C, D & G SYS | 15 | 35 | 50 |
| CCC-7-SF<br>TACTICAL MANEUVERS | 4 | 40 | 44 |
| CCC-9-SF<br>FLAGHOIST SIGNAL PROCEDURES | 11 | 41 | 52 |
| **Total CCC** | 227 | 779 | 1006 |
| **FSO** | | | |
| FSO-M-10-SF<br>SMOKE INHALATION | 11 | 28 | 39 |
| FSO-M-11-SF<br>BURNS | 10 | 29 | 39 |
| FSO-M-1-SF<br>BTL DRESSING STATIONS | 11 | 27 | 38 |
| FSO-M-2-SF<br>PERS CASUALTY TRANSPORT | 17 | 26 | 43 |
| FSO-M-3-SF<br>COMPOUND FRACTURES | 14 | 28 | 42 |
| FSO-M-4-SF<br>SUCKING CHEST WOUND | 10 | 26 | 36 |
| FSO-M-5-SF<br>ABDOMINAL WOUNDS | 10 | 25 | 35 |
| FSO-M-6-SF<br>AMPUTATION | 11 | 26 | 37 |
| FSO-M-7-SF<br>FACIAL WOUND | 14 | 26 | 40 |
| FSO-M-8-SF<br>ELECTRIC SHOCK | 15 | 26 | 41 |
| FSO-M-9-SF<br>MASS CASUALTY | 2 | 13 | 15 |
| **Total FSO** | 125 | 280 | 405 |
| **INT** | | | |
| INT-1-SF(MS)<br>INTEL COLL & RPTG TEAM | 7 | 38 | 45 |

**Table C.1—Continued**

| Mission Area | Where Exercise Was Completed | | |
| | In Port | Under Way | Total |
| --- | --- | --- | --- |
| INT-1-SF(OP) | | | |
| OPINTEL DATA COLL | 8 | 41 | 49 |
| INT-1-SF(RP) | | | |
| INTEL RPTNG - LOCATORS | 7 | 40 | 47 |
| INT-2-SF(OP) | | | |
| OPINTEL PLOT & BRIEF | 8 | 30 | 38 |
| INT-2-SF(RP) | | | |
| INTEL INFO RPTS | 5 | 37 | 42 |
| INT-3-SF(OP) | | | |
| C2W/INFO WARFARE | 1 | 14 | 15 |
| INT-4-SF(RP) | | | |
| SURVINTCOLEX | 1 | 11 | 12 |
| INT-5-SF(RP) | | | |
| INCSEA/DNGR MIL | 3 | 8 | 11 |
| **Total INT** | 40 | 219 | 259 |
| **MOB–Engineering** | | | |
| MBGGM– | | | |
| CLASS B FIRE GTG | 8 | 14 | 22 |
| MBGTM– | | | |
| CLASS B FIRE GTM MODUL | 3 | 31 | 34 |
| MCASF–GT | | | |
| COOL AIR FAILURE | 4 | 31 | 35 |
| MCBF–B | | | |
| FIRE IN MAIN SPACE | 19 | 31 | 50 |
| MCCFG– | | | |
| CLASS C FIRE GEN | 5 | 20 | 25 |
| MCCFS– | | | |
| CLASS C FIRE SWB | 8 | 21 | 29 |
| MCFED– | | | |
| CLASS C FIRE EDS | 10 | 38 | 48 |
| MEPTV–PT | | | |
| VIBS HI GTM | 6 | 26 | 32 |
| MGGOS– | | | |
| GAS GEN OVERSPEED | 1 | 23 | 24 |
| MGGS– | | | |
| GAS GEN STALL GTM | 3 | 23 | 26 |

**Table C.1—Continued**

| Mission Area | Where Exercise Was Completed | | |
| --- | --- | --- | --- |
| | In Port | Under Way | Total |
| MGHIT– HIGH GT INLET TEMP | 6 | 21 | 27 |
| MHBGTG– HOT BEARING GTG | 3 | 21 | 24 |
| MHTIT– PT INLET TEMP HI GTM | 1 | 30 | 31 |
| MLCWS– LOSS CHILL WATER | 3 | 10 | 13 |
| MLEPC– LOSS OF EPCC | 6 | 21 | 27 |
| MLFOP– LOSS MN ENG F/O PRESS | 14 | 28 | 42 |
| MLGGO– GTG LOW L/O PRESS | 3 | 19 | 22 |
| MLPACC– LOSS OF PACC CONSOLE | 3 | 16 | 19 |
| MLPTO– LOW LOP GTM | 1 | 29 | 30 |
| MLSC– LOSS STEERING CONTROL | 6 | 30 | 36 |
| MLSCU– LOSS SHAFT CONTRL UNIT | 3 | 17 | 20 |
| MMF– FLOODING IN MAIN SPACE | 10 | 27 | 37 |
| MMFOL– MAJOR FUEL OIL LEAK | 17 | 31 | 48 |
| MNVGG– GTG NOISE/VIBRATION | 6 | 32 | 38 |
| MOSGG– OVERSPEED SSGTG | 4 | 8 | 12 |
| MPSFG– GTG POST SHUTDOWN FIRE | 5 | 30 | 35 |
| MPSFP– POST SHUTDOWN FIRE | 2 | 41 | 43 |
| MPTOS– PWR TURB OVERSPEED | 5 | 20 | 25 |

## Table C.1—Continued

| Mission Area | Where Exercise Was Completed | | |
|---|---|---|---|
| | In Port | Under Way | Total |
| MNVRG–<br>NOISE/VIB MRG/SHAFT | 4 | 24 | 28 |
| **Total MOB–Engineering** | 169 | 713 | 882 |
| **MOB–Damage Control** | | | |
| MOB-D-10-SF<br>RESCUE/ASSIST | 7 | 5 | 12 |
| MOB-D-11-SF<br>SETTING MATL COND: PHASE 1: YOKE,<br>PHASE 2: ZEBRA | 35 | 65 | 100 |
| MOB-D-12-SF<br>U/W HULL DAMAGE PH 1 AND 2 | 13 | 22 | 35 |
| MOB-D-13-SF<br>SHORING | 8 | 21 | 29 |
| MOB-D-14-SF<br>FIRE EXTINGUISHING/SMOKE<br>CLEARING | 29 | 36 | 65 |
| MOB-D-15-SF<br>CHEMICAL ATTACK | 7 | 2 | 9 |
| MOB-D-20-SF<br>ISOLATE/PATCH DAMAGED PIPE | 14 | 29 | 43 |
| MOB-D-21-SF<br>MAJOR FLOOD MAIN PROPULSION<br>SPACE | 7 | 19 | 26 |
| MOB-D-23-SF<br>LOCATING DC FITTINGS | 25 | 37 | 62 |
| MOB-D-24-SF<br>DARKEN SHIP | 7 | 76 | 83 |
| MOB-D-27-SF<br>HELO CRASH F/F | 13 | 22 | 35 |
| MOB-D-2-SF<br>RELIEF VITAL STATIONS | 10 | 34 | 44 |
| MOB-D-31-SF<br>TOXIC GAS DRILL | 5 | 3 | 8 |
| MOB-D-3-SF<br>MANNING BAT STATIONS | 33 | 61 | 94 |
| MOB-D-4-SF<br>EMERG INTERIOR COMM | 8 | 17 | 25 |

## Table C.1—Continued

| Mission Area | Where Exercise Was Completed | | |
| --- | --- | --- | --- |
| | In Port | Under Way | Total |
| MOB-D-5-SF<br>TOPSIDE DAMAGE | 8 | 18 | 26 |
| MOB-D-7-SF<br>PROV CASUALTY POWER | 3 | 7 | 10 |
| MOB-D-8-SF<br>MAJOR CONFLAGRATION | 4 | 11 | 15 |
| MOB-D-9-SF<br>MAIN PROPULSION FIRE | 20 | 37 | 57 |
| **Total MOB–Damage Control** | 256 | 522 | 778 |
| **MOB–Seamanship** | | | |
| MOB-S-14-SF<br>SAREX | 5 | 4 | 9 |
| MOB-S-2-SF<br>HEAVY WEAX | 1 | 11 | 12 |
| MOB-S-33-SF<br>HOIST/LOWER BOATS | 23 | 60 | 83 |
| MOB-S-34-SF<br>RESCUE SWIMMER | 1 | 22 | 23 |
| MOB-S-7-SF<br>PREPS-ABANDON SHIP | 5 | 13 | 18 |
| **Total MOB–Seamanship** | 35 | 110 | 145 |
| **NCO** | | | |
| NCO-11-SF<br>COMBAT CLASS C FIRE | 18 | 28 | 46 |
| NCO-12-SF<br>EQUIPMENT CASUALTY REPAIR | 9 | 29 | 38 |
| NCO-15-SF<br>ALTERNATE PWR SOURCE | 8 | 15 | 23 |
| NCO-16-SF<br>ECC/ESS | 6 | 9 | 15 |
| NCO-18-SF<br>SECURITY DRILLS | 55 | 7 | 62 |
| NCO-19-SF<br>SMALL ARMS QUALS | 39 | 24 | 63 |
| NCO-1-SF<br>PREP FOR ELEC SPACES | 8 | 22 | 30 |

## Table C.1—Continued

| Mission Area | Where Exercise Was Completed | | |
| --- | --- | --- | --- |
| | In Port | Under Way | Total |
| NCO-28-SF ROE | 18 | 21 | 39 |
| NCO-29-SF DEFENSE VS U/W SWMRS | 30 | 2 | 32 |
| NCO-30-SF SHIP PENETRATION-BASIC | 35 | 6 | 41 |
| NCO-32-SF TERRORIST A/C ATTACK | 15 | 12 | 27 |
| NCO-33-SF SMALL BOAT ATTACK | 33 | 16 | 49 |
| NCO-34-SF BOMB THREAT | 37 | 3 | 40 |
| NCO-35-SF HOSTAGE SITUATION | 25 | 2 | 27 |
| NCO-36-SF FLOATING DEVICES | 17 | 10 | 27 |
| NCO-38-SF VBSS | 8 | 19 | 27 |
| NCO-39-SF ATFP PLANNING P/S | 10 | 8 | 18 |
| NCO-40-SF ATFP PLAN EXEC P/S | 10 | 7 | 17 |
| NCO-41-SF ATFP PLANNING W/S | 7 | 8 | 15 |
| NCO-42-SF ATFP PLAN EXEC W/S | 8 | 6 | 14 |
| **Total NCO** | 396 | 254 | 650 |
| **STW** | | | |
| STW-1-SF MISSION DATA UPDATE | 25 | 45 | 70 |
| STW-21-A TLAM/C LAUNCH (SIM) | 22 | 29 | 51 |
| **Total STW** | 47 | 74 | 121 |
| **SUW** | | | |
| SUW-18-SF DATABASE MGMT | 15 | 53 | 68 |

**Table C.1—Continued**

| Mission Area | Where Exercise Was Completed | | |
| --- | --- | --- | --- |
| | In Port | Under Way | Total |
| SUW-3-I SUW FREEPLAY EX | 1 | 16 | 17 |
| SUW-9-SF SURF TRKG NTDS/AEGIS | 4 | 48 | 52 |
| SLAMEX | 23 | 32 | 55 |
| **Total SUW** | 43 | 149 | 192 |
| **Grand Total** | 1,459 | 3,666 | 5,125 |

# BIBLIOGRAPHY

ASD FM&P—*see* Assistant Secretary of Defense, Force Management and Personnel.

Assistant Secretary of Defense, Force Management and Personnel, *Training Simulators and Devices*, Washington, D.C.: U.S. Department of Defense, Directive 1430.13, August 22, 1986.

Brown, David, "Naval Training: The New Era," *Navy Times*, September 2, 2002, p. 14.

Burlage, John, "Training Shake-Up: Top-to-Bottom Reorganization Aims to Streamline Schools, Meet Fleet Needs," *Navy Times*, July 8, 2002, p. 8.

Chief of the Maritime Staff (Canada), *Maritime Command Combat Readiness Requirement*s, March 30, 2001.

Chief of Naval Operations, Training and Education Assessment Division (N813), *Surface Simulation*, March 29, 2002.

Chief of Naval Operations and Commandant, U.S. Coast Guard, *Universal Navy Task List (UNTL)*, OPNAV Instruction 3500.38A/ USCG COMDT Instruction M3500.1A.

COMFLTFORCOM—*see* Commander U.S. Fleet Forces Command.

COMNAVSURFLANT—*see* Commander Naval Surface Force U.S. Atlantic Fleet.

Commander Naval Surface Force U.S Atlantic Fleet, Surface Force Training Manual Bulletins, Washington, D.C.: Department of the

Navy, COMNAVSURFLANT/COMNAVSURFPAC Instruction 3502.3 CH-3, October 20, 1992.

_____, *Surface Force Training Manual*, Washington, D.C.: Department of the Navy, COMNAVSURFLANT/COMNAVSURF-PAC Instruction 3502.2E, December 17, 1999.

Commander, Naval Surface Force, Atlantic, Afloat Training Group, Atlantic (ATGLANT), *Mission, Functions and Tasks (MFT)*, COMNAVSURFLANT Instruction 5450.8, May 29, 2001.

Commander Naval Surface Forces, *Surface Force Training Manual*, Washington, D.C.: Department of the Navy, COMNAVSURFOR Instruction 3502.1, February 27, 2002.

Commander Operational Test and Evaluation Force, *Use of Modeling and Simulation (M&S) in Operational Testing*, Norfolk, Va.: Department of the Navy, COMOPTEVFORNIST 5000.1, September 5, 1995.

Commander, Second Fleet, Battle Group Inport Exercise–East Letter of Instruction, naval message, June 2002.

Commander, Third Fleet, *PC EMPSKED Employment Terms/Fuel Factors*, Washington, D.C.: Department of the Navy, Third Fleet Instruction 9261.1, September 28, 1998.

COMTHIRDFLT—*see* Commander, Third Fleet.

Commander U.S. Fleet Forces Command, *Fleet Forces Command Modeling and Simulation Project Team (M&S FPT) Charter*, Washington, D.C.: Department of the Navy, COMFLTFORCOM Instruction 3502.1, March 6, 2002a.

Commander U.S. Fleet Forces Command, *Fleet Forces Command Fleet Training Strategy*, Washington, D.C.: Department of the Navy, COMFLTFORCOM Instruction 3501.3, May 28, 2002b.

Department of the Navy, Office of the Secretary, *Department of the Navy Modeling and Simulation Management*, Washington, D.C., SECNAV Instruction 5200.38A, February 28, 2002.

Department of the Navy, Office of the Secretary, *Verification, Validation, and Accreditation (VV&A) of Models and Simulations*, Washington, D.C., SECNAV Instruction 5200.40, April 19, 1999.

DoD—*see* U.S. Department of Defense.

Dorsey, Jack, "Technology May Be Key for New Navy Training," *Norfolk Virginian-Pilot*, September 23, 2002.

Foreman, Ross, "Training to Go High-Tech: Great Lakes Sets Plans in Motion for Simulator," *Chicago Tribune*, October 29, 2002.

InsideDefense.com, "Atlantic Fleet to Begin Training with Virtual Technology Next Month," *Inside the Navy*, October 14, 2002.

International Maritime Organization, *Standards of Training, Certification and Watchkeeping for Seafarers (STCW 95)*, 1978m as amended in 1995 and 1997 (STCW Convention) and Seafarer's Training, Certification and Watchkeeping Code (STCW Code) including Resolution 2 of the 1995 STCW Conference, as amended in 1997), London, 1996.

Kapos Associates, Inc., *Assessment of the Utility of Simulation in Fleet Training*, Arlington, Va., TR 2-98, January 19, 1998.

Kaufman, Gail, and Amy Svitak, "Transforming Training: DoD Creates Requirements, Acquisition Strategies," *Defense News*, July 1–7, 2002, p. 1.

Miller, Sandy, "CNO Visits Navy Simulation, Task Force EXCEL Hub in Orlando," *Navy Newsstand*, November 20, 2001. Online at http://www.new.navy.mil/search/display.asp?story_id=2763

National Research Council, Naval Studies Board, *Technology for the United States Navy and Marine Corps, 2000–2035*, Vol. 9: Modeling and Simulation, Washington, D.C.: National Academy of Sciences, 1997.

Naval Research Advisory Committee, *Modeling and Simulation*, November 1994.

"MSI Simulator Training," *Marine Log*, Vol. 106, No. 7, July 1, 2001, p. 27.

"The Next Generation of Ship-Handling Simulators," *Sea Power*, Vol. 44, No. 2, February 1, 2001, pp. 41–43.

Office of the Chief of Naval Operations, Status of Resources and Training System (SORTS), Washington, D.C.: Department of the Navy, Naval Warfare Publication 1-03.3, September 1987.

Office of the Chief of Naval Operations, *Projected Operational Environment (POE) and Required Operational Capabilities (ROC) for DDG-51 (Arleigh Burke) Class Guided Missile Destroyers* (U), Washington, D.C.: Department of the Navy, OPNAV Instruction C3501.311, July 8, 1994.

Office of the Chief of Naval Operations, Space, Information Warfare, Command and Control Directorate Navy Modeling and Simulation Management Office, *Navy Modeling and Simulation Master Plan*, Washington, D.C.: Department of the Navy, February 21, 1997a.

Office of the Chief of Naval Operations, *Fleet Exercise Publication (FXP) 3 for Strike Warfare (STW), Surface Warfare (SUW), Intelligence (INT), Command and Control Warfare (C2W), and Command, Control and Communication*, Rev. G, Washington, D.C.: Department of the Navy, March 1997b.

Office of the Chief of Naval Operations, *Navy Training System Requirements, Acquisition, and Management*, Washington, D.C.: Department of the Navy, OPNAV Instruction 1500.76, July 21, 1998.

Office of the Chief of Naval Operations, *Surface Warfare Training Strategy*, Washington, D.C.: Department of the Navy, OPNAV Instruction 1500.57A, August 3, 1999.

Office of the Under Secretary of Defense, Acquisition, Technology, and Logistics, Report of the Defense Science Board Task Force on Training Superiority and Training Surprise, Washington, D.C.: U.S. Department of Defense, January 2001.

OPNAV—See Office of the Chief of Naval Operations.

Schank, John F., Harry J. Thie, Clifford M. Graf II, Joseph Beel, and Jerry Sollinger, *Finding the Right Balance: Simulator and Live*

*Training for Navy Units*, Santa Monica, Calif.: RAND Corporation, MR-1441-Navy, 2002.

Shearon, Blane T., "The Cost Effectiveness of West Coast Distributed Simulation Training for the Pacific Fleet," Monterey, Calif.: Naval Postgraduate School, December 2001.

Silkman, William R., Jr., "Simulator Utilization for Navy Training," Newport, R.I.: Naval War College, July 19, 2001.

SURFTRAMAN—See Commander Naval Surface Force U.S. Atlantic Fleet and Commander Naval Surface Forces.

Systems Simulation Branch (N81) Engineering Environment Division Combat Systems Department Naval Surface Warfare Center Dahlgren Division, AEGIS Combat Trainer System (ACTS) MK 50 User's Manual, March 2000.

U.S. Department of Defense, *DoD Modeling and Simulation (M&S) Management*, Washington, D.C., DoD Directive 5000.59, January 4, 1994.

U.S. Department of Defense, Under Secretary of Defense, Acquisition and Technology, *Modeling and Simulation (M&S) Master Plan*, Washington, D.C., DoD Directive 5000.59-P, October 1995.

U.S. Department of Defense, *DoD Modeling and Simulation (M&S) Verification, Validation, and Accreditation (VV&A)*, Washington, D.C., DoD Instruction 5000.61, April 29, 1996.

U.S. Department of Defense, Under Secretary of Defense, Acquisition Technology, *DoD Modeling and Simulation (M&S) Glossary*, Washington, D.C., DoD Directive 5000.59-M, January 1998.

Young, Christopher, "The STCW Convention: A Handbook of Highlights," handout, 2002. Also online at http://www.uscg.mil/stcw/s-handbk1.htm (October 6, 2003).